D0025301

IEE POWER SERIES 19

Series Editors: Professor A. T. Johns
J. R. Platts

ELECTRICAL SAFETY
a guide to the causes and prevention of electrical hazards

Other volumes in this series:

Volume 1 **Power circuits breaker theory and design** C. H. Flurscheim (Editor)
Volume 2 **Electric fuses** A. Wright and P. G. Newbery
Volume 3 **Z-transform electromagnetic transient analysis in high-voltage networks** W. Derek Humpage
Volume 4 **Industrial microwave heating** A. C. Metaxas and R. J. Meredith
Volume 5 **Power system economics** T. W. Berrie
Volume 6 **High voltage direct current transmission** J. Arrillaga
Volume 7 **Insulators for high voltages** J. S. T. Looms
Volume 8 **Variable frequency AC motor drive systems** D. Finney
Volume 9 **Electricity distribution network design** E. Lakervi and E. J. Holmes
Volume 10 **SF_6 switchgear** H. M. Ryan and G. R. Jones
Volume 11 **Conduction and induction heating** E. J. Davies
Volume 12 **Overvoltage protection of low-voltage systems** P. Hasse
Volume 13 **Statistical techniques for high-voltage engineering** W. Hauschild and W. Mosch
Volume 14 **Uninterruptible power supplies** J. D. St. Aubyn and J. Platts (Editors)
Volume 15 **Principles of digital protection** A. T. Johns and S. K. Salman
Volume 16 **Electricity economics and planning** T. W. Berrie
Volume 17 **High voltage engineering and testing** H. M. Ryan (Editor)
Volume 18 **Vacuum switchgear** A. Greenwood

ELECTRICAL SAFETY
a guide to the causes and prevention of electrical hazards

J. Maxwell Adams

The Institution of Electrical Engineers

Published by: The Institution of Electrical Engineers,
London, United Kingdom

Reprinted 1997, 2004

The Institution of Electrical Engineers,
Michael Faraday House,
Six Hills Way, Stevenage,
Herts. SG1 2AY, United Kingdom

British Library Cataloguing in Publication Data

A CIP catalogue record for this book
is available from the British Library

ISBN 0 85296 806 X

Printed and bound by Antony Rowe Ltd, Eastbourne
Produced by Bookchase (UK) Ltd

Contents

1 Physiological effects of electric current **1**

2 Electric power systems **7**
2.1 Low voltage overload and short-circuit protection 7
2.2 Earth fault protection 15

3 The philosophies of earthing **18**
3.1 Options 18
3.2 Earthing electrical enclosures 20
3.3 Earth return circuit 22
3.4 The need for bonding 25
3.5 Limitations 26
3.6 Reducing the risks 28
3.7 Electricity supply systems—principles and practice 31
3.8 Characteristics of supply 33
3.9 Earthing systems 37
3.10 Measurement of electrode contact resistance and ground resistance 38
3.11 Review 40
3.12 Bibliography 41

4 Cables and fires **42**
4.1 Polyvinyl chloride cables and wiring 44
4.2 Mineral insulated (MI) cable 46
4.3 Silicone rubber cables and wiring 47
4.4 Cross-linked polyethylene cables 47
4.5 Thermal ratings 48
4.6 Bibliography 51

5 Electrical equipment for use in explosive atmospheres **52**
5.1 Development of area classification 53
5.2 Assessment of extent of zones 55
5.3 Other properties of explosive atmospheres 58
5.4 Electrical sources of ignition 59
5.5 The design of electrical apparatus for use in hazardous areas 61

5.6 General principles of design 61
5.7 Harmonised standards and the New Approach 64

6 Protection by flameproof enclosure 66
6.1 Principles of testing 68
6.2 Permitted flameproof gaps 69
6.3 Methods of connecting flameproof equipment 73
6.4 Applications and limitations of flameproof protection 75
6.5 Review 78
6.6 Bibliography 78

7 Protection by intrinsic safety 79
7.1 The design of intrinsically safe systems 80
7.2 Applications of IS systems 82
7.3 Testing for intrinsic safety 85
7.4 Ignition by overheated components 88
7.5 Bibliography 89

8 Electrical apparatus in areas subject to flammable dusts 91
8.1 Dust clouds 91
8.2 Deposited dust 94
8.3 Surface temperatures 94
8.4 Design of apparatus 95
8.5 Causes of dust fires and explosions 98
8.6 Bibliography 98

9 Design, workmanship and maintenance 99
9.1 Designing for safety 100
9.2 Workmanship and maintenance 103

10 Stored energy 108
10.1 Batteries 108
10.2 Capacitors 118
10.3 Bibliography 121

11 Electric welding 122
11.1 Arc welding 122
11.2 Resistance welding 124
11.3 Hazards associated with electric welding 125
11.4 Review 127
11.5 Bibliography 127

12 Lightning phenomena and protection 128
12.1 The nature of lightning 128
12.2 Development and characteristics of a lightning stroke to the ground 130
12.3 Protection of buildings and services 133
12.4 Earthing of buildings and lightning conductors 135
12.5 Protection of tank farms 136
12.6 Summary of lightning protection 138

12.7 Statistical risks 138
12.8 Bibliography 139

13 Coping with static **140**
13.1 Electric charges on solid surfaces 140
13.2 Electric charges on powders 148
13.3 Electric charges in liquids 150
13.4 Electric charges in gases 153
13.5 Ignition of explosive gases 153
13.6 Electrostatic painting and finishing 154
13.7 Bibliography 156

14 Electromagnetic radiation **157**
14.1 Circuits which produce and accept radiation 161
14.2 Electromagnetic interference 163
14.3 Electromagnetic pulse 170
14.4 Radiological effects 175
14.5 Bibliography 177

15 Earth currents and their effects **178**
15.1 Electric traction 178
15.2 Railway signalling 182
15.3 Corrosion of buried structures 185
15.4 Cathodic protection 187
15.5 Bibliography 190

Index **191**

Chapter 1

Physiological effects of electric current

> 'On the 12th of August 1897, I took a pass at the Kennington underground railway for the monument, and was lowered in the lift. On asking where I was to go, I was informed "straight ahead". The platform being badly lighted, I fell on the line, with trains due every four minutes. In attempting to rise I grasped the conductor, and had some 400 volts of electricity playing through me, experiencing the sensation as if a cannon-ball were rushing through me from head to foot, and I lost consciousness, feeling death inevitable. I recovered somewhat on being lifted on the platform by two of the porters, who dressed my wounds at the lavatory, having sustained a severe cut in the forehead, an elbow bared to the bone, and a tendon slightly burnt between thumb and finger, notwithstanding which I experienced no headache nor other inconvenience; and was soon enabled to travel home; and, although now in my eighty-third year, am happily free from rheumatism, to which I was before subject. Possibly, if my heart had been weak, the shock would have killed me. But I am under the impression that electricity judiciously administered, is beneficial in acute cases of rheumatism and nervous complaints.'

The above extract is from a symposium of the Institution of Electrical Engineers, held in London on 27 February 1902, on the subject of electric shock.

Published records of accidents in the eleven years from 1880 to 1890 showed that in the whole world only few people had been killed by electric shock. Yet twelve years later, at an Institution of Electrical Engineers meeting on the subject in London, speaker after speaker described the nature and effects of electric shock from both observed and first-hand experience.

At that time electric power supplies were rapidly being extended—mainly for lighting— following the opening of the first power station in England by Edison in 1882, and the use of Ferranti's high voltage alternating current system a few years later. Legislation to afford protection to employees liable to be subjected to 'electric shock' was incorporated in the Workshop and Factories Act of 1901.

Hazards from the relatively new and fast-growing use of electricity was thus an important and emotive subject. It was generally believed that systems below about 500 volts were not particularly dangerous and could be treated with little respect. Only the higher voltages were considered lethal, execution by electrocution having been made Law in the United States in 1889.

Experience seemed to show that the effects of electric shock were less severe if the victim were totally relaxed at the time and there was much anecdotal evidence regarding persons who had been drunk, asleep or unconscious. Those of low mental intellect were also believed to be less susceptibile to electrocution.

The author of one of the papers presented at the 1902 IEE symposium had made numerous personal tests at various voltages and under conditions that today would be regarded as extremely foolhardy. His presentation, entitled 'Electric shocks at 500 volts' could have been subtitled 'Familiarity breeds contempt', as the following excerpt indicates:

> 'To grasp with two bare hands, two pieces of metal at 500 volts would give a very painful shock, but a light and quick touch is no worse than the shock from a half-pint Leyden jar, an experience more familiar to schoolboys than to engineers. A 500 volt shock may be described as worse than touching a kettle of boiling water, not as bad as touching a red-hot poker, about the same as touching a soldering iron at working heat, or as when an inexperienced blacksmith's boy picks up a black-hot horseshoe. These shocks are common incidents in the daily work of a careless linesman; nearly all of those who are practically engaged in electric traction work receive more shocks than they like, but they agree that they might reduce the number by taking more care.'

It was known that the most critical organ in the case of electric shock was the heart and it was accordingly assumed that those with any cardiac problems were more vulnerable. In fact it was suggested that anyone intending to become an electrical engineer should first undergo a cardiac examination, and if any weakness were discovered, should be advised to follow some other profession where there was no chance of an electric shock from handling apparatus.

The clinical effects of electricity passing through the body were beginning to be understood, with various proposed methods of resuscitation in the case of total collapse. Artificial means to stimulate the heart, such as the hypodermic injection of ether, brandy etc. were suggested. It was even recommended that high voltage sub-stations and generating stations should be equipped with a hypodermic syringe, together with small phials of the necessary drugs.

The effect of spasm preventing one from letting go was widely recognised at that time and another author of a paper to the IEE suggested that all live metal should be of such a shape that it could not be grasped with the hands, as the danger lay in getting the hands closed round the metal and not being able to release oneself. He added 'I have personally known a man held by 100 volts, due to his hands contracting around metal. He used terribly bad language, but could not get away'.

A contributor to the above discussions was one of His Majesty's Factory Inspectors who referred to five general principles contained in the 1901 Factory Act:

(1) Metal which is not part of the circuit should be earthed
(2) Live metal should be protected with insulating material
(3) Work should not be done on metal at high voltage except by skilled men in cases of urgent necessity

(4) When work has to be done on high voltage metal, proper safeguards such as gloves and insulated mats should be supplied
(5) Artificial respiration should be properly understood.

It is interesting to note that these basic tenets, embodied into English Law nearly a century ago, are virtually identical to present-day regulations, except that much lower voltages are now considered to be dangerous.

We can see that the problems arising from the use of electric power were realised and well understood very soon after it came into use as a source of energy. Chief among these were its interference with the telegraph systems of the time and the effects of electric shock.

Current passing through the body can produce four types of physiological damage, or trauma. Firstly, there is the possibility of skin burns and necrosis of the underlying tissue. The degree of burning will depend upon the energy liberated at the point of contact and its duration. That is to say

$$I^2Rt \quad \text{or} \quad (E^2/R)t$$

where I is the current flowing, R is the resistance at the point or area of contact, and E is not the source voltage, but the drop in voltage between the live conductor and the surface of the body.

Thus, with a very dry skin, a high voltage shock could produce a severe burn without necessarily electrocuting the victim. A lower voltage applied to a wet or sweaty skin could, however, cause death without any evidence of burning, particularly if the path of the current is across the chest. For obvious reasons most deaths due to electricity occur at consumer voltages and in these cases the only visible injury is typically a slight discolouring of the skin where a live part has been grasped.

From an engineering point of view, the human body is in effect a tank of electrolyte, the tank itself (the skin) having a relatively high ohmic resistance. Most of the burning will therefore be generated on the surface of the body at the point of contact.

Skin resistance can normally only be deduced (on a live subject) from the overall resistance between two places where the electric current enters and leaves the body. Tests have shown that this value can range from a few hundred ohm to several thousand ohm.* It is also found that the resistance is far higher at low voltages. Electrical safety is therefore greatly enhanced by reducing the source potential, since the current flowing will be more than proportional to the applied voltage.

With any electric shock the greatest internal current density will generally be along the shortest path between the two areas of contact—most frequently between one hand and the feet or between the two hands. The density will be greatest in the immediate vicinity of the point of contact and the current may destroy tissue under what appears to be a superficial skin burn. Careful medical attention is necessary here, since a life-threatening infarct may subsequently arise, i.e. an area of dead tissue due to the blocking of the bloodstream which normally nourishes it.

The second type of trauma is by paralysis of the breathing centre of the brain.

*Exceptionally low values will be present if part of the body is immersed in a liquid electrode—bath water for example.

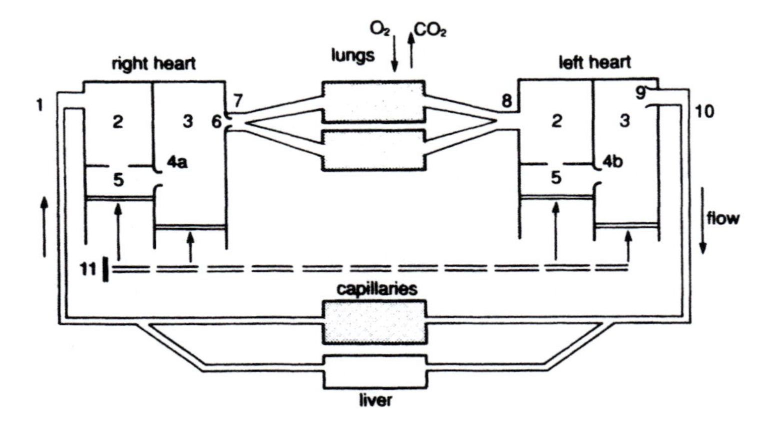

Figure 1.1 Schematic diagram of the function of the heart

1 Vena cava (carrying blood from capillaries and liver to the heart via the veins)
2 Right and left atria (reception vessels)
3 Right and left ventricles (main pumps)
4a Tricuspid valve (non-return valve)
4b Mitral valve (non-return valve)
5 Right and left auricles (priming pumps)
6 Pulmonary arterial valve (non-return valve)
7 Pulmonary arteries (supply to the lungs)
8 Pulmonary veins (supply from lungs to heart)
9 Aortic valve (non-return valve)
10 Aorta (main discharge line from the heart to the body)
11 Heart muscle or myocardium (pump drive)

This of course leads to deoxygenation of the blood and is equivalent to suffocation. In this condition the pulse may also be irregular, feeble or nonexistent.

It is vitally important that some form of resuscitation is applied. Movement of the lungs will also provide a massaging effect on the heart encouraging it to beat correctly. Patients who have been rendered lifeless by an electric shock can often be revived by artificial respiration which should be applied as quickly as possible and maintained for not less than an hour before the case can be regarded as hopeless. Any method is better than none since even minor amounts of cardiopulmonary activity can be sufficient to sustain life.

The two other types of trauma concern the heart itself and are therefore likely to be lethal unless some sort of pulse can be rapidly restored. Failure of the heart to deliver oxygenated blood to the head will result in irreversible brain damage and then death within a very short time. One cannot survive more than a few minutes if the brain is deprived of arterial blood. In mechanical terms the human heart is a compact 280 g assembly, comprising two separate 2-stage displacement pumps working in series and in synchronism. It has a continuous rating of about 4·5 W and a short-time rating of at least 20 W. Its 'specification' requires it to be

- Self-powered by extraction of energy from the pumped medium (i.e. by using some of the oxygen from the blood)
- Completely maintenance-free

- Capable of continuous operation at 60–74 strokes/min for at least 600 000 h with 100% reliability
- Self-regulating by nervous control from the brain.

Delivery: 5 litres/min (short-time rating 20–30 litres/min)
Discharge Pressure: 0·18 kg/cm^2 (2·5 psi)
Discharge Rate: approx. 0·45 m/s

The right pump takes in the vitiated blood from the tissue capillaries and from the liver and discharges it to the two lungs for reprocessing (removing CO_2 and adding O_2). The left pump takes in the restored blood from the lungs and circulates it to the tissue capillaries via the arteries. Figure 1.1 shows the heart function schematically. Figure 1.2 shows the components of the heart and medical terminology.

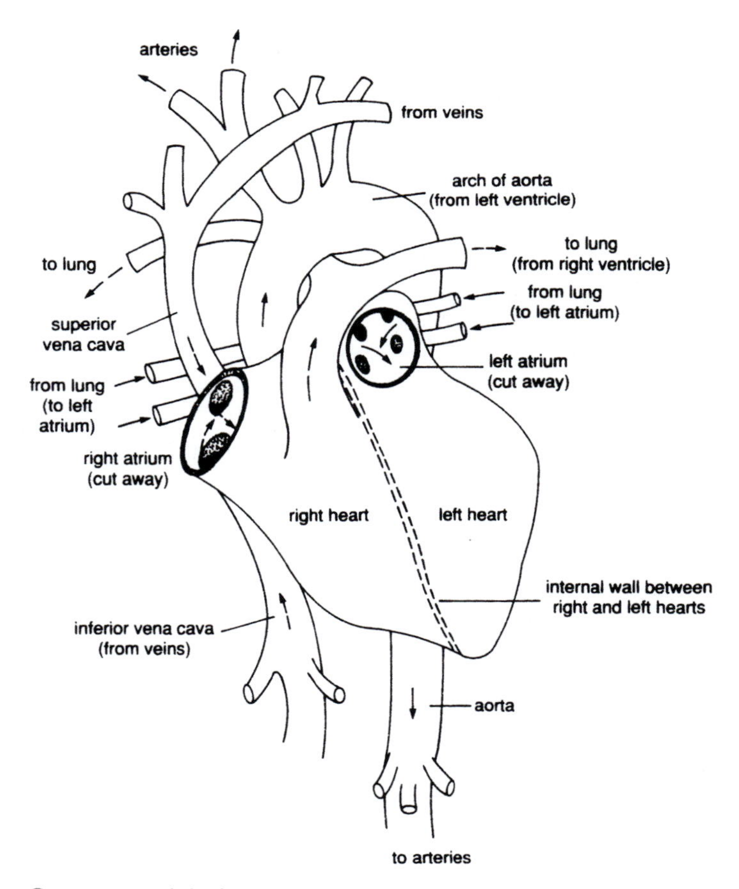

Figure 1.2 Components of the heart

Both atria contract at the same time followed by the contraction of the two ventricles, forcing blood into the lungs and into the aorta, respectively.
solid arrows: oxygenated blood flow
broken arrows: deoxygenated blood flow

Although operating as a pulsation pump, the intermittent discharge from the heart becomes a constant flow in the capillaries due to the elasticity of the arteries combined with the non-return valves in the circuit.

As mentioned, there are two ways in which the heart can be upset by the passage of an electric current: it will either stop completely, producing a state of cardiac arrest, or the synchronous contractions of the auricles and ventricles (see Figure 1.1) will cease to be co-ordinated. In this state, the little bundles of muscle surrounding the heart cease to contract in unison and start to tremble or twitch so that no effective pumping action is produced. The resulting condition is known as cardiac fibrillation or, more generally, ventricular fibrillation since it is the major cavities which are primarily affected. Unless the condition can be quickly rectified death will ensue and it is generally considered that ventricular fibrillation is the most common cause of death in electric shock.

Where no burning has occurred, survivors of electric shocks generally recover completely: a very small number have suffered after-effects attributed to their experience. These include cataracts, angina and various nervous disorders. Numbness and pain sometimes persist for a long time after the initial shock.

The foregoing comments refer primarily to low voltage shocks from single-phase domestic supplies. The situation is quite different if contact is made at higher voltages—it is not even necessary to make physical contact. The national press has this week reported the death of a boy playing on a railway footbridge: he was thrown through the air after dangling some kitchen foil at the power line carrying 25 kV. A few years ago a child had to have a leg amputated and an arm removed at the shoulder after he had waved a scaffold pole near an 11 kV line while standing on a flat roof. Power lines look passive and harmless but they show no mercy when approached too closely. Many certified picnic areas and caravan sites have high voltage power lines passing overhead and kite flying on these sites is an obvious hazard.

Here in the East Anglian prairies, agro-industry uses very large equipment for spraying crops and digging drains and dykes. Pole-mounted HV lines are sometimes caught by these machines without being broken or grounded. Electrocution is then most likely—and happens—as the driver steps out of his cab which is then at line potential. In these circumstances, drivers would be well advised to jump out rather than step out.

Bibliography

PD6519: 1995 Parts 1 and 2, Effects of current on human beings and livestock. BSI

Chapter 2

Electric power systems

2.1 Low voltage overload and short-circuit protection

The present internationally agreed definition of low voltage is a supply pressure normally exceeding extra-low voltage but not exceeding 1000 V a.c. or 1500 V d.c. between conductors, or 600 V a.c. and 900 V d.c. to earth. Extra-low voltage is defined as a supply pressure normally not exceeding 50 V a.c. or 120 V ripple free d.c., whether between conductors or to earth.

This definition applies specifically to separated extra-low voltage systems (SELVs). These are supply sources electrically isolated from earth and from other sources of potential.

2.1.1 Overloads

The term overload is used here to differentiate from overcurrent, which can include excessive earth leakage, as discussed in Chapter 3. By overloads we are referring merely to abnormally high currents—but not short-circuit currents—flowing through the normal load circuits. In this sense there are basically two causes of overload. The first affects only cabling and wiring and is a result of either connecting too many appliances to one circuit, e.g. supplying a washing machine with a 3 kW immersion heater from the same 3 kW socket as an electric fire, or of operating an appliance which takes a normal current above the circuit rating, e.g. running a 6 amp electric kettle from a 5 amp lighting socket.

The second cause of overload—which may affect the cabling and wiring, the apparatus or all three—is due to the reduced impedance of a connected load. This commonly arises from the stalling or slowing down of a motor, so reducing its back emf. Stalling can be due to the motor load, such as a pump, or a machine tool seizing up while running. It can equally be a failure to start because the initial torque required by the load is greater than the accelerating torque available from the motor. Mechanical overload, causing the machine to slow down, will also be reflected as an overload in the electrical demand from the driving motor. For example, crude oil used as a fuel in power stations has to be heated to reduce its viscosity before being pumped to the burners. If the temperature is allowed to fall, the additional load on the fuel pumps may cause the motors to trip on overload.

Transformers and generators can also be overloaded by reducing the impedance of their connected loads. In the case of motors and most transformers the protective device will be upstream of its protégé whereas generators have to be protected downstream, as shown in diagrammatic form in Figure 2.1.

Most types of equipment can alternatively be safeguarded from dangerous or harmful overheating by thermal detectors within their windings. For all but the smallest pieces of apparatus these embedded temperature detectors (ETDs) need to operate through an external tripping relay or contactor. For single-phase fractional horsepower motors, as used in their millions for domestic equipment, ETDs can function by direct switching of the motor supply. Some of these devices contain minute fusible elements; others operate by means of a bistable, bimetallic microswitch that will self-reset when the winding has cooled.

Referring again to Figure 2.1, when an overload occurs the voltage should be disconnected at a point as close as possible to the cause in order to minimise the extent of the loss of supply. Only one overload trip should operate—the one nearest to the problem. This delegation of function is generally referred to as discrimination or grading. It is an important factor in the overall design of systems where there is a sequence of protective devices in series.

Overload protection is concerned with defending wiring, cables and apparatus windings from temperatures capable of causing permanent damage or undue weakening of the insulation. Accordingly, when selecting overload devices, two other factors need to be considered besides the question of adequate discrimination:

- The thermal inertia of the conductors concerned
- The maximum permissible temperature excursion of the respective insulation

An external tripping device may be triggered by a bimetallic helix heated by the load current, or a part of it, and should preferably have a similar heating and cooling time constant to the protected circuit. For apparatus such as an electric motor, it is generally impractical for the overload heater to be a thermal model

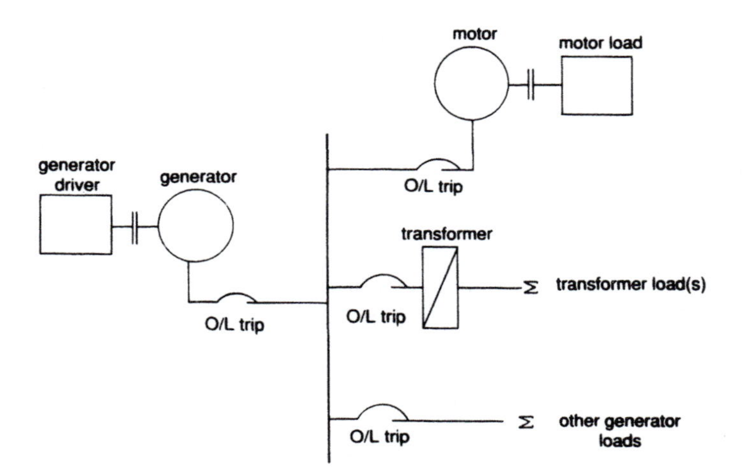

Figure 2.1 Schematic diagram showing the relative positions of overload/fault protection devices for generator, motor, transformer and other sub-circuit loads

of the windings embedded in an iron core. A thermal trip normally has a cooling time constant of less than a minute. Although this can be extended by fixing it to a mass of metal, it will not reach the cooling time of a large motor, which may be a matter of hours. In most cases thermal trips will therefore heat up and cool down far more quickly than the windings they are protecting.

A motor with a long starting period—such as occurs with high inertia or high starting torque loads—is normally designed for a maximum frequency of restarts from the full load heated condition—up to 2 starts/hour from hot is often specified. Figure 2.2 shows the heating effect on the motor and on a thermal overload device for a rapid succession of on-load starts. As can be seen, the windings can be overheated in this way without tripping the overload device because the motor has a longer cooling time constant than the overload heater.

Different ambient air temperatures between starter and motor may also have

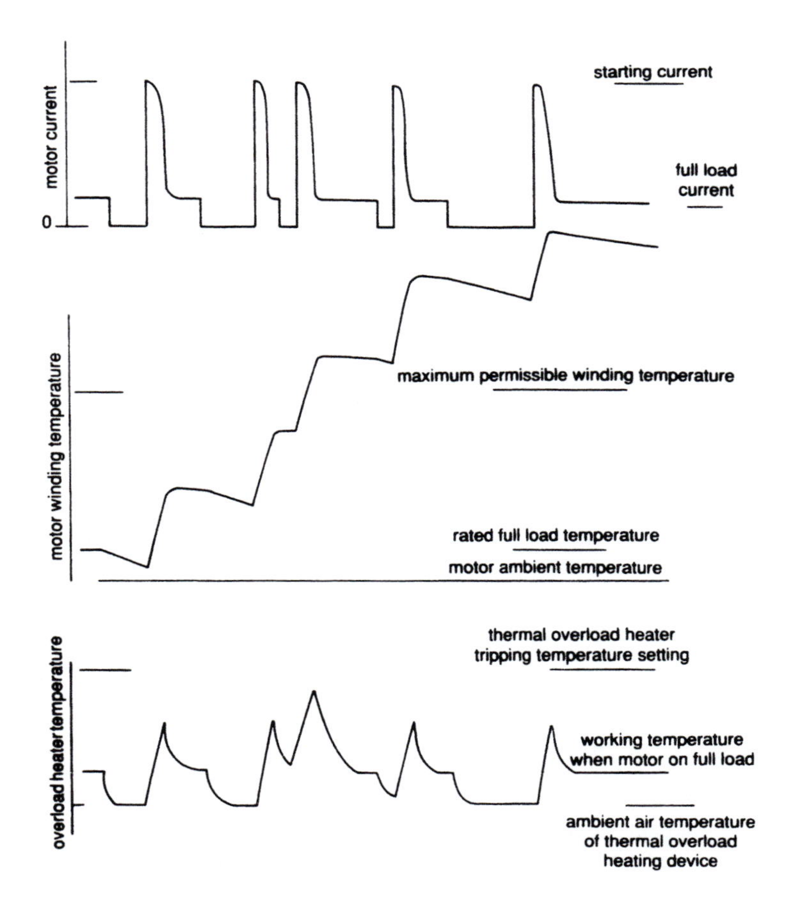

Figure 2.2 Motor winding temperatures and thermal overload trip temperatures for frequently repeated starting duties. Typical curves for a large motor with high inertia load—shown to the same time scales (x-axis)

an adverse effect. A thermal overload device has an inverse time/current tripping characteristic which should be calibrated at the ambient temperature in which it is to be used. In large installations the motor control centre (MCC) will be in a separate room, generally maintained at about 15°C. If the ambient temperature is below the calibration ambient the response will be less sensitive (see Figure 2.2). Conversely, a motor may overheat when the ambient temperature is above its rated air temperature. Most totally enclosed and totally enclosed fan cooled (TEFC) machines are rated for a maximum air temperature of 40°C and water/air cooled machines for a maximum cooling air temperature of 25°C. Higher ambients are present in many applications. Fan motors mounted above steam boilers, for instance, may have to work in an air temperature which is continuously above 70°C. This does not preclude the use of thermal trips, but does mean that the motor windings need to be insulated with materials suitable for these higher working temperatures.

Lack of co-ordination due to different time constants and ambient temperatures does not arise when the overload protection is by embedded temperature detectors, since these respond only to the actual temperature of the windings in which they are embedded. For larger machines, it is now common practice to incorporate one or more devices known as a thermistor, which is an abbreviation or acronym for 'thermally sensitive resistor'. These devices consist of doped sintered oxides and, being semi-conductors, can have a wide range of resistance/temperature characteristics. The first thermistors were developed in the 1940s and had resistances which sharply decreased with temperature, albeit over a narrow temperature range. They are known as negative temperature coefficient (NTC) thermistors. More recently, materials with a positive characteristic have been developed and it is the PTC types which are more commonly used as ETDs in machine windings. Because the steeply rising part of the curve—which approximates to an S shape—occurs only over a limited temperature range, it is necessary to select a formulation which will respond at the most critical temperature for the application.

By direct detection of the winding temperature it is possible to keep a machine in operation even under limited fault conditions. An important example of a limited fault condition is the loss of voltage on one phase of the supply to a three-phase cage motor: the machine will continue to run but with reduced torque and increased currents in the windings. The alternative method of protection against overheating due to loss of a phase is to install a starter relay which will trip the whole supply to the motor immediately one phase is lost. This may not be necessary if the motor is running on part load at the time. It is important to note, however, that the standard motor starter with overload and short-circuit protection will not necessarily safeguard the motor windings from overheating when voltage is present on only two of the three phases.

Overheating of three-phase induction motors is also possible if its torque/speed curve is not suitable for the torque/speed characteristic of the load. Figure 2.3 shows a typical cage induction motor torque curve with two possible load curves relating to a high pressure pump. For load curve *b*, the pump non-return valve is held closed until the motor approaches its running speed. The motor then runs up to its normal speed and takes full-load current. With load curve *a*, the non-return valve is allowed to open as soon as the pump delivery pressure exceeds the back pressure in the pump discharge line. In this case the motor torque is

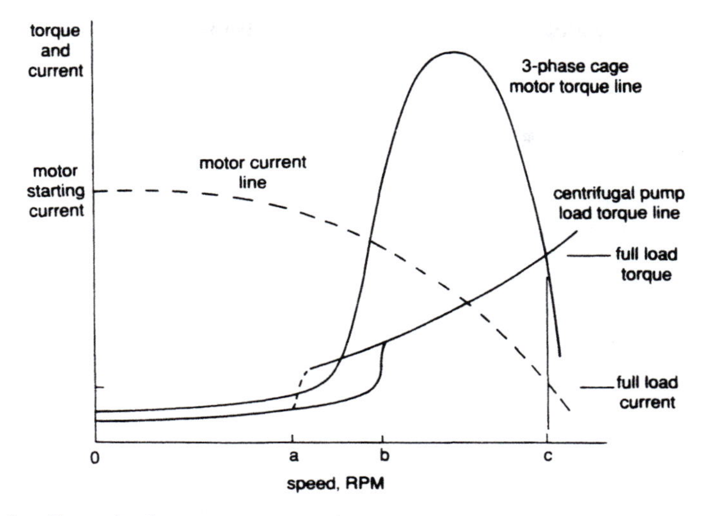

Figure 2.3 Example of a pump motor overload due to the pump delivery valve opening too early

When non-return valve opens at *a*, load torque increases above motor torque and motor ceases to accelerate further, while drawing overload current. When NR valve opens at *b*, motor is able to run up to normal full load speed *c*

not sufficient to accelerate beyond the first intersection between motor and load characteristics. The motor then stays at this speed, taking an overload current.

Machine windings can also be protected by electromagnetic tripping devices. A suitable time delay is then generally necessary to prevent loss of supply due to starting currents and switching surges. A time lag can be ensured using an inductive disc, an oil dashpot or a mechanical escape mechanism, all of which can have an automatic instantaneous reset feature.

The simplest form of protection is the traditional fuse. It has a number of limitations but is capable of supplying both overload and short-circuit protection. Consideration of fuses therefore follows a brief analysis of short-circuit protection.

The thermal inertia of wires and cables is far less than that of machine windings, where the iron core provides a considerable heat sink. Hence, overload protection devices for cable should not in principle have a significant time lag. However, insulating materials used for wiring and the primary insulation of cables are normally allocated two limiting temperatures:

- Maximum continuous running temperature
- Maximum short-term withstand temperature.

This enables a device with a time delay or an inverse time characteristic to be used for the overload protection of cables and wiring as well as machines and appliances.

The insulation of machine and transformer windings is internationally classified according to its thermal stability. British Standard BS 2757 1994, which corresponds to IEC Publ. 85, specifies seven temperature classes, each having a different maximum continuous working temperature at which a reasonable length of life can be expected. In each class various permissible materials are listed as possible insulants. No such classification exists for cables, each insulating material used having a defined temperature limitation. The temperatures for some typical cable and winding materials are shown in Table 2.1.

The limited choice of permissible insulation temperatures, plus the discrete number of machine frame sizes and cable sizes, usually prevent either cables or apparatus being utilised to their full design capability if overheating is to be avoided under all foreseen conditions. Cables will also in many instances have to be oversized in order to reduce the voltage drop.

It has to be recognised that overload protection by analogue devices is always a compromise which raises many difficulties in the economic design of safe systems. But although the temperatures of windings and appliances can be directly determined, this is scarcely practicable for cables in distribution circuits, particularly where thermal conditions vary along the route of the cable.

2.1.2 Short-circuits

The rapid clearance of short-circuits is invariably necessary to prevent overheating of cables and wiring feeding the fault, and it is important to remove the source of energy from the fault itself which may set fire to nearby flammable materials.

Electrocution and even electric shocks are not hazards associated with line to neutral or line to line faults, but they are associated with line to earth faults. These are discussed in some detail in the following chapter on the philosophy of earthing.

A short-circuit in a cable or in equipment is nearly always accompanied by an earth fault. Line to neutral or line to line contact produces vaporised metal, leading to a conducting ionised zone in the vicinity. This will in most cases offer a low impedance path to earthed metal. In fact, any short-circuit fault in Class I equipment, i.e. equipment surrounded by an earthed enclosure, can be expected to become an earth fault. Large currents can then flow in exposed metal creating alarming or even dangerous potential differences.

So for reasons of shock as well as the possibility of smoke and fire and the rapid overheating of the circuit upstream of the fault, the supply should be disconnected as quickly as possible. The circuit has to be broken while there is an abnormally high current flowing and system design must ensure that the short-circuit protection device has a breaking capacity at least equal to the task. To specify the required

Table 2.1 Limiting temperatures for some insulating substances commonly used in cables and windings

Insulant	Continuous running temperature (°C)	Maximum short-time withstand temperature (°C)
General purpose pvc	70	160
Impregnated paper	80	160
Cross-linked polyethylene	90	250
Class B winding insulation	130	195
Class F winding insulation	155	220
Class H winding insulation	180	245

rupturing capability, the maximum prospective short-circuit current (PSCC) must first be determined. This can be a complicated procedure requiring information about the impedance of the transformers and supply cables concerned and the contributory effect of any large motors in the system. The fault level, calculated on a worst case basis, will be invalidated if the supply network is subsequently reinforced, for instance with an additional transformer.

The rupturing capacity of a circuit-breaker, fuse or motor starter contactor may be specified in terms of megaVolt-amps (MVA) for high voltage circuits and in kiloamps for low voltage circuits. The ability of the protective device to clear any short-circuit fault is usually taken for granted, but if it proves unable to interrupt the supply under a fault condition, the result can be quite serious. Firstly, the device may explode violently with the emission of incandescent metal. Secondly, adjacent circuit-breakers or other protective equipment may also be damaged and thirdly, a more senior upstream device will, hopefully, trip the supply with consequent loss of power to other circuits.

To avoid the need for many differently rated rupturing capacities, low voltage switchboards and motor control centres may be designed for a PSCC of either 50 000 amps or 16 000 amps, while miniature fuses for sub-circuits and individual items of equipment are standardised to clear prospective faults of 4000 or 1000 amps. At the higher short-circuit currents, the force between conductors also needs to be taken into account in switchboard and MCC design.

2.1.3 Protection by fuses

The inclusion of a fusible link in a power supply is the oldest and simplest method of protection against short-circuit faults. Fuses are also extensively used for overload protection, which in many respects is more onerous than the disconnection of short-circuit faults. Their advantages are that they are relatively cheap and have no moving parts to stick or wear. However, they do have the following limitations compared with more sophisticated devices:

2.1.3.1 They are not reusable

When a fuse blows, a replacement cartridge or a suitable fuse wire has to be found and fitted—possibly by candle- or torch-light—in order to restore the circuit. In a distribution board, if there are no tell-tale signs on the fuses, it may be necessary to pull out each fuse holder to find the fuse which has to be replaced or rewired.

2.1.3.2 They may permit blown fuses to be replaced with larger sizes

Many of the smaller fuses and rewirable fuse holders can accept current links of various ratings. There is always a temptation to put in a larger fuse if a particular circuit repeatedly blows its fuse. Spare, larger circuit-breakers are not usually to hand and there is no easy way to change them when they trip. Increasing the fusing current will deny protection from an already faulty or overloaded circuit.

Some high rupturing capacity (HRC) cartridge fuses and fuse holders are designed so that only fuses with the same characteristics can be fitted. On troublesome circuits, replacing one of these fuses a few times soon costs as much as a miniature circuit-breaker.

2.1.3.3 Discrimination between fuses in series is not always precise

True discrimination depends on whether the junior or minor, downstream fuse completely interrupts the fault current before the senior or major, back-up fuse starts to melt. In practice, the major fuse may not become open-circuit but could be critically weakened by partially melting. The apparent initial discrimination may then fail completely on a later occasion. A current ratio of at least 2:1 is needed for reliable discrimination.

2.1.3.4 They are not particularly suitable for protection of 3-phase induction motors

Where the supply is via three fuses, only one may blow, leaving the machine running on two phases as a single-phase motor. This can cause overheating and burnout of the windings unless a so-called single-phasing relay is included in the circuit to the motor.

2.1.3.5 They are not suitable for precise overload protection of cables

The accepted rule for cable overloads is that the circuit should be interrupted within 4 hours if the current remains 45% above the continuous rating of the cable, e.g. a cable installed where it can take 30 amps continuously should be protected by a device which will interrupt the circuit within 4 hours at a current of

$$30 \times 1{\cdot}45 \text{ A} = 43{\cdot}5 \text{ A}$$

Since the temperature rise of a cable is approximately proportional to the square of the current, this rule permits the normal full-load temperature rise of a cable to be increased by 110% ($1{\cdot}45^2 = 2{\cdot}10$) for up to 4 hours. Most fuses are designed to blow in 4 hours at 145% of their nominal current rating. This means that a cable able to take 30 A continuously can be protected by a 30 A fuse. Fuses complying with British Standards 1361 and 1362 and general purpose fuses complying with BS 88 Parts 2 and 6 meet this requirement, but BS 3036 (rewirable) fuses need to have a nominal current rating of only 72·5% of the cable rating, as they are less sensitive to overload currents.

The snag with all this is that, at 45% overload, the shape of the fuse current/fusing-time curve is almost asymptotic to the time axis. In other words, if it takes 4 hours to blow at 45% over-current, a fuse will probably withstand 40% over-current indefinitely. Fuses are, therefore, incapable of defending cables against prolonged overloads of less than about 145% of the continuous rating of the cable. Circuit-breakers, on the other hand, can be designed with a combined thermal and electromagnetic feature which can more closely protect cables under various fault conditions.

In spite of the above limitations, fuses are still used in the majority of single-phase low voltage circuits, although their place in the market is being increasingly taken over by miniature and moulded case circuit-breakers. Rewirable fuses are used in the United Kingdom at up to 30 amps for 250 volt circuits and have a rated breaking capacity of up to 4000 amps (1 MVA). Cartridge fuses designed for specific voltages, currents and breaking capacities can be provided for full-load currents up to 1600 A and voltages up to 660 V. There are also many designs of high voltage fuses, normally pole-mounted for the isolation of overhead lines.

For short-circuit protection most fuses will blow in a time of about one millisecond, which is only some 18° of a 50 Hz a.c. cycle—not that this is of great benefit in a power circuit as the resulting current chop tends to produce an inductance kick, some of which is reflected into the supply network. The use of fuses for the overload protection of appliances and equipment raises the question of a suitable fusing factor, namely the ratio between the nominal carrying current of the fuse and its minimum melting current. Most cartridge fuses have a fusing factor of about 1·4 and the factor for rewirables is around 2·0. These fairly large ratios are necessary to ensure that the full-load current will not unduly heat the fusing element and cause it to deteriorate in service.

2.2 Earth fault protection

The currents which a single-phase fuse or circuit-breaker have to clear most frequently are line to earth fault currents. These will generally be far less than a line to neutral fault current and the corresponding times for some circuit-breakers to trip or a fuse to blow will be far longer than a millisecond. We are now looking at a situation where the fault current flows through exposed metal, creating accessible differences of potential. In a theoretically worst case, there is very little volt-drop in the line from the sub-station and most of the potential is consumed in the earth return circuit. Exposed metal at the fault is thereby raised nearly to line voltage representing a dangerous potential with respect to other earthed metal by-passed by the fault.

The actual potential between these exposed metal parts will depend on the resistance of the equipotential bonding in the vicinity of the fault and that of the earthing conductor between the fault and the bonding system. In any event, the occurrence of an earth fault is to be regarded as a dangerous condition which cannot be allowed to persist. By general international (IEC) agreement, earth faults need to be de-energised within specified time limits. Some of these are shown in Table 2.2, e.g. for circuits at 230 volts to earth there are two permissible clearance times of 5 and 0·4 seconds.

The electromagnetic tripping devices in circuit-breakers enable them to clear a fault in 0·1 seconds once the current has exceeded the set tripping value. Fuses, however, are essentially thermal devices and so have a form of inverse current/time characteristic. In either case, to ensure that the circuit is interrupted within the required time, the fault current has to exceed a specific value which will be limited by the impedance of the earth loop. Its value can be calculated from the ratio of voltage and earth loop impedance, that is

$$I_e = V_e/Z_s$$

where I_e is the earth fault current, V_e is the system voltage to earth and Z_s is the impedance of the earth loop.

The earth loop impedance is thus a critical factor which has to be co-ordinated with the current/time characteristic of the protective device to obtain the required speed of earth fault clearance.

To bring some of these abstractions into focus we can take the practical example of an electric iron being used near an electric cooker. The cooker can be assumed to be at earth potential. The iron will be supplied from a wall socket. There will

Table 2.2 Maximum disconnection times for earth faults on single-phase final circuits

System voltage (phase—earth)	Disconnection times (s)		
	For stationary equipment	For socket outlets and hand-held Class I apparatus with low resistance CPC-meb* connection	for socket outlets and hand-held Class I apparatus with standard CPC earth connection
120	5	5	0·8
230	5	5	0·4
400	5	5	0·2

*Connection between circuit protective conductor for the apparatus and the main equipotential bonding of the surrounding zone

be a flexible lead to the iron containing a line, a neutral and a third wire, known to the man on the Clapham omnibus as the green and yellow one and to the Chartered Electrical Engineer as the 'earth', while the time-served electrician calls it the CPC or circuit protective conductor. When there is an earth fault in the iron, current flows from the line, through the CPC back to the neutral or star point of the supply transformer. While this current is flowing there will be a potential difference between the body of the iron and the cooker, which is not taking the fault current and so remains at earth potential. According to the table above, the fault current causing the potential difference should be interrupted within 0·4 seconds. If the fault were on the electric cooker it could be allowed to persist for 5 seconds as it is not hand-held portable equipment.

If the electric iron were to have a low resistance equipotential zone bonding connection to the cooker there would be less chance of a dangerous potential occurring between the two items of exposed metal. In accordance with Table 2.2 it is then acceptable for the earth fault in the iron to persist for up to 5 seconds. These 5 and 0·4 second rules for earth faults dominate the design of single-phase electrical services and greatly complicate their specification and testing.

One of the critical factors arising from these rules is the maximum allowable distance from a socket outlet back to its fuse board. This has to be determined not only in consideration of the permissible volt-drop, but also of the maximum permissible earth loop impedance, either of which may be the critical parameter. Each type of protective device and current rating will interrupt the circuit in the specified times at different fault currents.

There are six classes of circuit-breaker and three classes of fuse, each with up to 14 current ratings, in common use in the UK. Table 2.3 shows, for 20 A ratings only, the currents necessary to produce 5 second and 0·4 second interruptions respectively. Column 3 of the table shows the corresponding maximum earth loop impedances for a 230 volt system.

Part of the earth loop impedance will be in the external earth return to the transformer neutral provided by the power supply company or authority. It is

*Table 2.3 Minimum currents necessary to cause disconnection within specified times for 20 amp protection devices and corresponding maximum permissible earth loop impedances at 230 volts**

Class of protective device	Current to trip within:		Corresponding earth loop impedance at 230 V to earth	
	5 s (amps)	0·4 s (amps)	5 s (ohms)	0·4 s (ohms)
Fuse type:				
BS 88 Part 2 and Part 6	79	130	2·91	1·77
BS 1361	82	135	2·81	1·70
BS 3036	60	130	3·84	1·77
Circuit-breakers:				
BS 3871 type 1	80	80	2·88	2·88
BS EN 60898 type B	100	100	2·30	2·30
BS 3871 type 2	140	140	1·64	1·64
BS 3871 type 3	200	200	1·15	1·15
BS EN 60898 type C	200	200	1·15	1·15
BS EN 60898 type D	200	400	1·15	0·57

*As a rule of thumb, these maximum impedances should be reduced by one third to allow for possible increase in the resistance of conductors due to the thermal effect of the earth fault current

therefore not always possible to meet the values shown in Table 2.3. In these circumstances another criterion is used: the potential difference between accessible exposed metal parts must not exceed 50 V for more than 5 seconds.

Under present standards, the design of apparently simple electrical services has become a complex procedure. In addition to protection from electric shock (which is almost entirely associated with earth fault protection) volt-drops, overloads, short-circuits, maximum demand and load diversity, ambient conditions and fault discrimination all have to be taken into account. Where motors are connected, starting currents also need to be considered. Rule of thumb tables and software for cable scheduling are available, but as discussed in the following chapter, the use of earth leakage sensing devices obviates many of the above calculations.

Chapter 3

The philosophies of earthing

3.1 Options

As we have seen, if a potential of more than about 50 V is applied across vital organs—the heart in particular—blood may cease to reach the brain, resulting in death by electrocution.

Because the senses cannot detect dangerous voltages per se except by touch, it is necessary to encase live conductors or to make them reasonably inaccessible. Nevertheless, the live rail of an electric railway, the bar of an electric fire and the socket of a table lamp are not encased nor particularly inaccessible, although all are potentially lethal. Such hazards have been accepted as part of the life-threatening aspects of so-called Western culture and, like crossing the road, broken glass and sweets from strangers, they fall within the syllabus of early safety training by parents.

In general, the live parts of domestic and industrial electrical equipment are enclosed to a varying extent in a protective casing. The degree of enclosure has been coded in an international (IEC) agreement. This is known as the IP Code (Ingress Protection) and is shown in Table 3.1. The IP Code differs significantly from the American NEMA Code and can be applied to both Class I equipment (i.e. equipment with exposed conducting parts) and to Class II equipment (i.e. double insulated, un-earthed enclosures).

Power and lighting sockets (also termed receptacles in America) have in the past been made with open holes, behind which live metal is accessible to an enterprising youngster, with a metal knitting needle for instance. Present UK designs to BS 1363 have shutters over the holes and the newer plugs are fitted with partially insulated live pins. However, these plug and socket assemblies still have disadvantages as regards safety.

In addition to safety requirements at the point of use, there are also certain conditions relating to the nature of the electricity supply.

In the United Kingdom, the electricity company has to connect its customers' supply system to earth at the source, that is to say at the sub-station. This fact is usually taken for granted because it is enforceable by law, but the necessity is not self-evident and does not always apply.

To examine the need to use an earthed source for domestic consumers let us assume the supply is floating, i.e. without an earth connection. Line and neutral then have no meaning; no shock can be obtained by grasping either of the two

*Table 3.1 Protection code for electrical apparatus**

The international (IEC) code for electrical enclosures is termed the IP Code. IP in this case does not indicate The Institute of Petroleum, which issues many codes of its own, but stands for Ingress Protection.

The degree of enclosure is denoted by the letters IP followed by two numbers to be interpreted as follows:

Ist No.	Protection against entry of solid objects
0	No protection
1	Entry up to 50 mm
2	Entry up to 12·5 mm
3	Entry up to 2·5 mm
4	Entry up to 1 mm
5	Dust proof
6	Dust tight

2nd No.	Protection against entry of water
0	No protection
1	Drip proof
2	Rain proof up to 15° from vertical
3	Rain proof up to 60° from vertical
4	Splash proof
5	Jet proof
6	Cascade proof
7	Proof against temporary immersion
8	Proof against permanent immersion

*A more detailed specification is given in BS EN 60529 (based on IEC Publ. 529:1989)

leads since there is no return path through the body. So far so good, but if a consumer inadvertently connects one of the two lines to earthed metal (in a faulty cooker for instance), the other line will be live to earth. Other consumers will not be aware of this and the situation will persist until a second consumer earths the other leg of the supply, at which point a fault current will flow and both consumers' supplies will trip. Such a situation is inherently dangerous. Furthermore a floating or all-insulated power system can be subjected to dangerous induced voltages. For these reasons it is illegal in the UK to supply consumers from an unearthed system. Note, we are speaking here of the earthing of the power supply itself, not of the earthing of the electrical enclosures which are considered in some detail in the following section.

However, if a consumer of electricity is using his own dedicated generator, as in the case of a motor car or a ship, the power supply does not necessarily need to be earthed at all. In a motor car, for reasons of cost, one leg of the battery is normally connected to the chassis which is then used as the return conductor. On ships, d.c. and single-phase a.c. are usually earthed at the centre point through two high resistances so that the first earth fault can be monitored and repaired, hopefully before a second earth fault on the other leg occurs and causes a short-circuit current to flow and trip the supply (see Figure 3.1).

Nowadays shipping generally has a 3-phase a.c. supply, but marine regulations permit either isolated or earthed neutral systems, except for tankers in which earthed systems are forbidden. An isolated system is to be preferred as an earth fault on an isolated system does not result in an outage or spark unless a second fault occurs. An important factor influencing the decision is that isolated installations are more subject to voltage spikes which can adversely affect solid state circuits and electronic components.

Earthing the neutral of a supply through a resistance or impedance will reduce the level of earth fault current and this practice is adopted in coal mines and many other fields, particularly overseas. Although reducing earth fault levels, it will not of course lessen the short-circuit fault breaking capacity (i.e. line to neutral or line to line) required for the protective equipment.

Some of the possible methods of earthing the power system are described in Section 3.8. Factors to be considered are continuity of supply, fault detection, fault levels, fault discrimination, safety and cost—not necessarily in that order.

The general practice is to achieve a limited level of safety by connecting one line to earth* at some point and the other line/s to each load via suitable fuses or circuit-breakers, which have the triple function of clearing earth faults, short-circuits and overloads.

3.2 Earthing electrical enclosures

The permanent connection of neutral to earth, either solidly or through an impedance, puts the phase line or lines at mains potential to earth. Consequently,

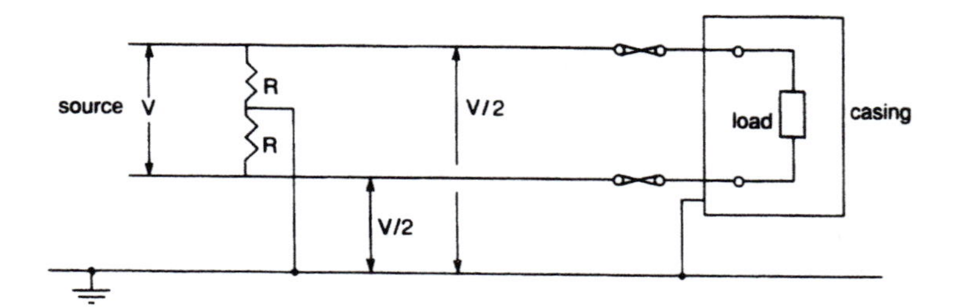

Figure 3.1 The use of centre point earthing to reduce the maximum possible voltage to earth to half the system voltage

Centre point earthing of d.c. and single-phase a.c. supply. With no earth fault present, direct contact voltage to earth is half the load voltage. No outage will occur until there is an earth fault on both legs. The monitoring and earthing resistors can be lamps or relays

*At the turn of the century, an earth was legally defined as follows: The earth is an earth, but an earth need not necessarily be the earth. Today, it is defined as: The conductive mass of the Earth, whose electric potential at any point is conventionally taken as zero.

all conducting casings of mains-fed apparatus and of cable screens etc. must themselves also be connected to earth. If they are left 'unearthed' they could remain at a dangerous potential indefinitely due to an internal fault since no fault current would flow. The system of having an earthed power supply and earthed enclosures is intended to ensure that any internal contact between the live conductors and the casing will produce a current high enough to operate the protective device. However, it should be noted that contact between the neutral and the casing can occur and persist without altering the current flowing.

The three possible situations are shown diagrammatically in Figure 3.2.

In the first instance, the supply which has one line earthed at the source is connected into equipment in an unearthed metal casing. This provides personal protection from direct contact with live conductors, but an accidental internal contact of the 230 V line to the housing leaves the whole enclosure at a dangerous potential to earth.

In the second instance, the casing is earthed and the hazard is rapidly removed by the earth fault current blowing the fuse in the 230 V line.

In the third instance there is no immediate hazard from the internal connection of the neutral to the casing, but the load current can flow undetected through exposed metalwork.

In principle, this earthed line/fused line system of supply is a fault-throwing

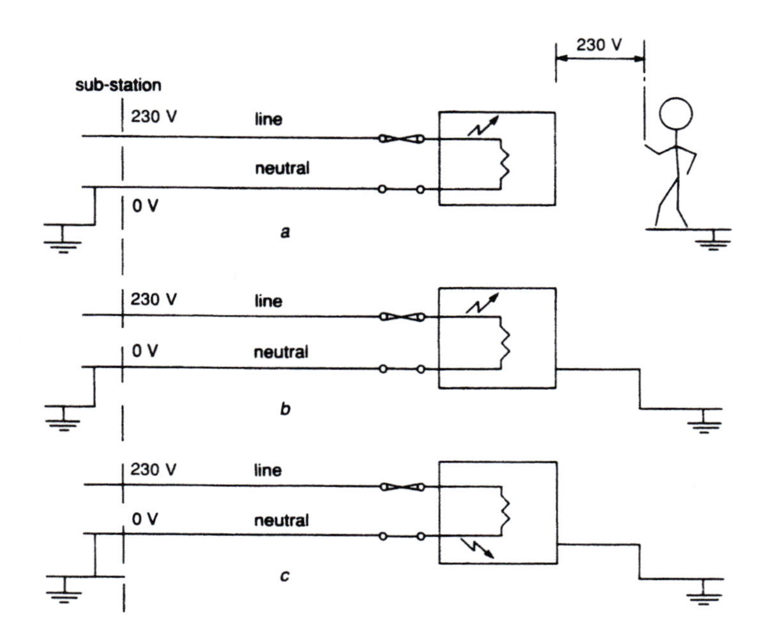

Figure 3.2 Effects of earth faults

a Unearthed casing remains live to earth
b Earthed casing enables fault current to blow fuse
c Neutral fault to earth remains undetected

or crowbar circuit* intended to ensure that an earth fault will create sufficient over-current to operate the fuse or circuit-breaker upstream of the fault. In order to achieve this, the return path for the fault current through the earth (instead of through the neutral line) must have a suitably low impedance. Furthermore, the protective device must be sensitive enough to clear the fault before the wiring and cabling overheats, but not so sensitive that false tripping occurs, due to a motor starting current for example.

3.3 Earth return circuit

While the earth fault current is flowing there will be a volt drop between the fault and the nearest earthing point of the neutral. This raises the voltage to earth of the metalwork near the fault and so is a possible hazard until the protection device opens the circuit. As previously noted, the voltage feeding into the fault should automatically be tripped off within 5 seconds and within 0·4 seconds for hand-held equipment. That is to say, a socket outlet must clear an earth fault within 0·4 seconds. These figures apply for supply voltages up to 277 V to earth.

Such a system will not of course be 'fail safe' and its reliability will decrease with time. In practice, the protective device carrying the earth fault current may not remove the source of energy in the required time (or at all) for any of the following reasons:

(a) A replacement fuse for a higher current, or a circuit-breaker with a higher setting may have been inserted, so that the protection is insufficiently sensitive.
(b) The impedance of the earth loop circuit may be too high, e.g. due to the drying out of the soil, or a high resistance connection.
(c) The fault itself may have a significant resistance, for instance, it may be an arcing fault so that the fault current is reduced due to the volt drop in the arc.
(d) The earth fault in the equipment may be away from the line end of the load so that the voltage to earth at the fault is reduced.
(e) The earth wire may have been omitted; it can also have been broken or disconnected at some point in the circuit accidentally, without any outward indication that the earth loop is incomplete.

These conditions are shown diagrammatically in Figure 3.3 as the requirements for protection by means of earth fault current.

All the possibilities described above point towards the probability that a high proportion of consumer installations will not respond correctly to an earth fault. The risk increases in time and it is undoubtedly the increasing fallibility of this simple method of protection which makes old wiring installations dangerous, particularly as regards risk of fire.

It should be emphasised here that far more people are killed in fires attributed to electrical equipment and wiring than are killed as a result of electric shock. The official figures for the United Kingdom (given in Table 3.2) indicate that

---

*A crowbar circuit is a term used in electronics to indicate a protective circuit which in the event of a fault that could damage equipment, will apply a short-circuit across the supply, causing the fuse to blow or a cut-out to operate, thus isolating the equipment from the supply.

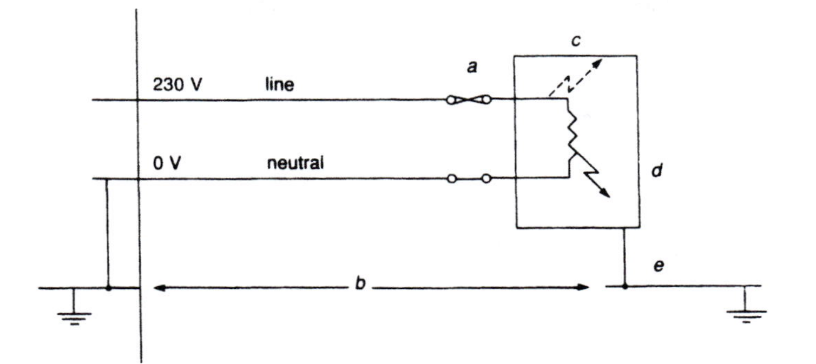

Figure 3.3 Conditions required for protection by means of earth fault current

a The correct fuse or breaker is used
b Earth loop has a sufficiently low impedance
c No effective impedance in the fault itself
d The fault is not too close to the neutral end of the load
e All metal round live conductors is earthed

between four and six times as many people die in fires caused by electrical installations as from electrocution.

The wiring codes of practice such as the British Standard 7671:1992 'Requirements for electrical installations. IEE Wiring Regulations. 16th Edn.' have attempted to overcome these ageing problems by specifying regular inspections. But these are costly to carry out and require a special instrument to measure the earth loop impedance, which is the most important critical parameter. In addition, a report on an old installation will often recommend very expensive corrective action, such as partial or total rewiring, to bring it up to present-day standards rather than because it is actually unsafe. As a result, the only time the electrical services in a flat or house are tested, or even inspected, is normally when the property changes hands.

Although safety standards continue to improve as regards domestic power supplies, they are little appreciated or understood by the average householder. This is largely unavoidable because of the ever increasing complexity of wiring standards. Familiarity with modern electrical practice is not encouraged by the

Table 3.2 Fatalities from fires caused by electrical installations, compared with fatalities by electrocution, in dwellings in the U.K. *

Year	Fires caused by electrical apparatus and wiring		No. of deaths due to electrocution
	No. of fires	No. of deaths	
1977	22 414	129	34
1978	24 641	189	28
1979	26 284	207	37

*Quoted by A.L. Kidd in the IEE Power Engineering Journal, Jan. 1987

technical words and phrases adopted by the regulations themselves—the meanings of which are sometimes at variance with everyday usage.

Metalwork surrounding electrical equipment is referred to in technical literature as an exposed-conductive-part. When these parts are made live by a fault—either with partial voltage elevation during the passage of the fault current, or with full voltage elevation because the casing has not been connected to earth—a shock received from the normally dead metalwork is termed 'indirect contact'. A shock received from the live conductor itself is termed 'direct contact'. The difference is shown in Figure 3.4.

The following methods of electrocution are evidently possible with no fault current flowing:

- Direct contact by grasping the line and the neutral or two phases of an energised supply. This is a very unlikely accident and effectively requires a willful attempt to kill oneself. The lethal current will not be detectable by any protective equipment since body resistance is fairly high and the current is indistinguishable from a healthy load current.
- Direct contact with a live conductor while part of the body is connected to earth. The current which flows through the body may well be lethal and sufficient to prevent one from letting go but will be insufficient to trip the supply. The simple crowbar circuit referred to above therefore offers no protection from death.
- Indirect contact with an unearthed casing which has an earth fault within it. Here again the standard circuit offers no protection from electrocution because the current through the body will not be sufficient to trip the supply.

As might be expected, the unearthed casing with an internal earth fault is the most common cause of electrocution. There are two reasons for this: firstly, the internal earth fault cannot be detected if the case is not earthed and secondly, the case is sitting indefinitely with a full mains voltage above other earthed parts such as service pipes, steelwork and concrete floors.

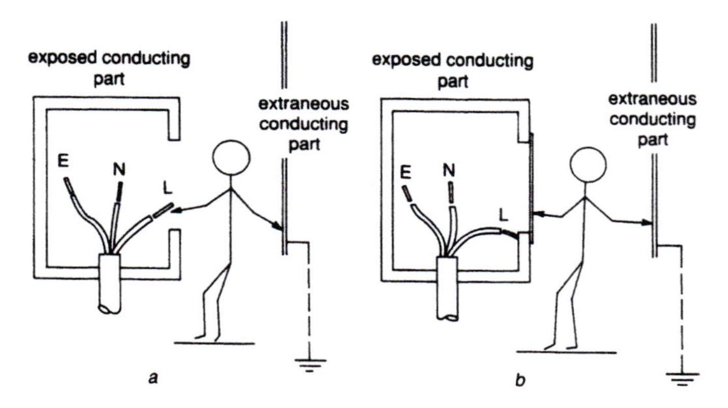

Figure 3.4 Direct and indirect contact

a Direct contact
b Indirect contact

3.4 The need for bonding

As mentioned, in national wiring regulations indirect contact has a special meaning, namely an electric shock (not just physical contact). Indirect contact can therefore only occur when an exposed-conducting-part is at a different potential from other adjacent metal or surfaces at earth potential, so that current will flow from the exposed-conducting-part to earth through the body. These other parts, not encasing electrical equipment, are termed extraneous-conducting-parts* (see Figure 3.4).

In order to prevent indirect contact (i.e. shocks from exposed-conducting-parts) all exposed-conducting-parts and extraneous-conducting-parts should be connected together with low resistance conductors to ensure they will all remain at the same potential. This is termed equipotential bonding and the area in question is then termed an equipotential zone.

In principle no bonding is required where there is no electrical equipment. However, it is normal practice to connect equipotential bonding throughout a building or factory since portable electrical equipment may be introduced. In many cases it is obligatory (see also Section 3.8).

It is inevitable that while a fault current is flowing through earthing and bonding conductors there will be some potential differences between adjacent metalwork due to volt drop. To minimise this, the sizes of the bonding and earthing conductors need to be proportional to the possible maximum fault current. Common practice is for the earthing conductor of the exposed-conducting-part to be at least half the size of the supply conductors and for the bonding conductors to be at least half the size of the earthing conductors, with certain limitations with regard to minimum and maximum sizes.

The assumption so far has been that with suitable equipotential bonding no dangerous potential differences can arise between conducting parts in the area, even when a fault current is flowing and accordingly there will be no danger from indirect contact.

Where there are semi-conducting floors which may be wet or dry, their surface resistance will be variable and indeterminate and to produce a true equipotential zone, e.g. in a cellar or milking parlour, is not really a practical proposition. How would one apply equipotential bonding to an old stone floor for instance?

This brings into focus a further question of exactly what needs to be bonded. The simplistic answer is that all accessible conducting material needs to be bonded and solidly earthed if it can present a return path from a live conductor back to the neutral. This sounds anomalous but by a return path we mean one with a resistance low enough to pass a dangerous current through the body.

Taking 2·5 mA as the maximum safe current through the body, Ohm's Law indicates that with a 230 V supply the minimum permissible resistance to the neutral is

$$230/0{\cdot}0025 = \text{about } 0{\cdot}1 \text{ Mohm}.$$

Hence, extraneous-conducting-parts need to be bonded if their resistance to the neutral is less than about 0·1 Mohm.

---

*The terms exposed-conducting-parts and extraneous-conducting-parts were first introduced into the 15th Edition of the IEE Wiring Regulations, but without the hyphens.

There are differences of opinion regarding the need to bond items such as metal office furniture, sliding patio doors and metal window frames. In some cases the bonding and earthing of metalwork actually increases the danger of lethal shock. For instance, to earth the frame of a hospital bed supported on anti-static rubber wheels of about one megohm resistance would greatly increase the danger of electrocuting the patient from faulty medical equipment.

A completely alternative solution to potential equalisation is to ensure that all extraneous-conducting-parts and exposed-conducting-parts are insulated from earth. This is then termed an 'earth-free' environment. Such areas must have a clear notice stating that it is an earth-free zone and that no earthed metal such as central-heating pipes, conducting flooring etc. may be introduced. This is a standard form of protection when live maintenance, repairs or testing have to be undertaken — as in radio repair workshops. The method is, of course, totally unsuitable for wet areas such as kitchens and bathrooms.

In principle there are thus three choices:

- to have an unearthed power supply and a normal 'earthy' environment
- to have an earthed power supply and an earth-free environment or
- to have an earthed power supply and an 'earthy' environment, with all extraneous conducting parts that may have a resistance to the transformer neutral of less than 0·1 Mohm bonded to any exposed conducting parts within arms reach. In addition, all the exposed conducting parts must have a suitably low earth loop impedance, as described in Section 2.2.

3.5 Limitations

The standard system of low voltage supply in the UK is now 230 V 50 Hz, single-phase or three-phase, fed from one or more delta/star connected transformers, with the secondaries connected to earth at the neutral points. Conducting enclosures and cable screens (exposed-conducting-parts) are also earthed and extraneous-conducting-parts are bonded together and earthed as necessary.

This system is intended to ensure that an earth fault will produce a sufficiently heavy current to trip the supply without producing dangerous potential differences. Although the method is widely used throughout the world, it is recognised that, by modern standards, additional safety features are needed in order to overcome the following shortcomings:

- Because the resistance of the human body is of the order of 1000 ohms, neither direct contact nor indirect contact will produce sufficient current to trip the supply, even when the current through the body exceeds the let-go value and causes electrocution. (230 V into 1000 ohms produces 230 mA which is lethal but is not a fault current capable of tripping a power supply.)
- The loss of an earth connection is not self-revealing and so can persist undetected. This is particularly dangerous in the case of hand-held apparatus fed from the mains by a flexible cable. Sudden tension on the flex at the plug

can disconnect the earth lead and allow it to touch the line terminal. The case of the hand-held apparatus is then live at 230 volts to earth. The side entry of the cable to the BS 1363 plug top is a disadvantage here compared with plugs in which the cable enters centrally so that tension on the cable will withdraw the plug (see Figure 3.5).

- When an earth fault occurs, the supply to the faulty part will not be automatically disconnected as intended if the impedance of the earth fault loop is too high. (In practice a high earth loop impedance is often more likely than an earth fault.)
- Earth leakage currents—as distinct from earth fault currents—can occur due to dampness, pollution or degraded insulation and are increasingly to be expected on older installations and equipment. Such leakage currents can persist and cause fires before sufficient current flows to trip the supply. (As already noted, fires caused by electricity kill more people than die from electric shock.)
- The neutral can become connected to earthed metal within a consumer's premises without producing any evident effect. The load current then returns through an unknown earth path which may or may not include the neutral in parallel. If there is no return path through the neutral, when the unknown return path is interrupted the metalwork on the supply side of the break will have a lethal potential of 230 V above earth (see Figure 3.6).
- If the source of power is switched to a small local generator, the voltage will

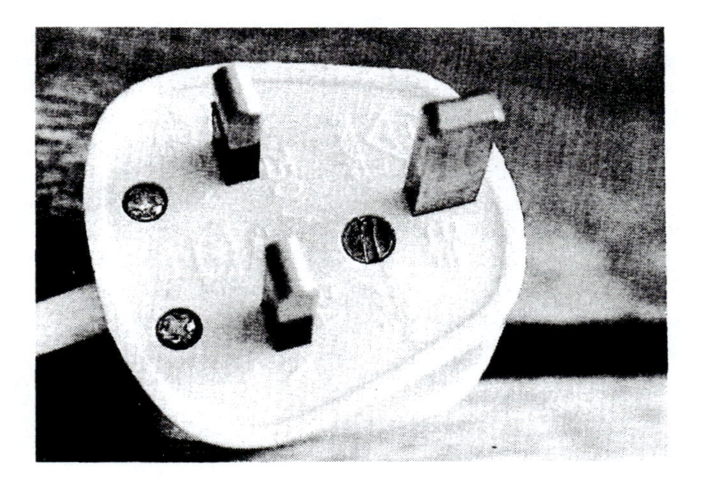

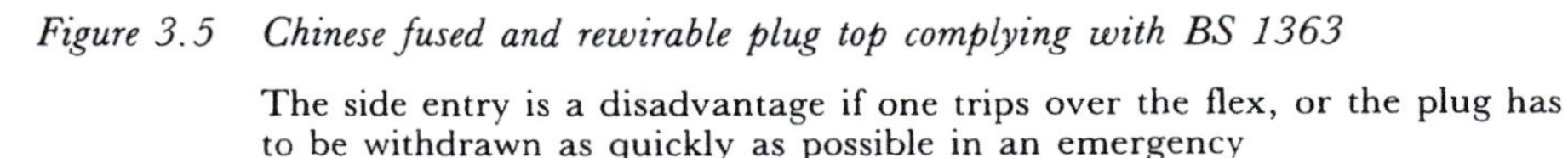

Figure 3.5 Chinese fused and rewirable plug top complying with BS 1363

The side entry is a disadvantage if one trips over the flex, or the plug has to be withdrawn as quickly as possible in an emergency

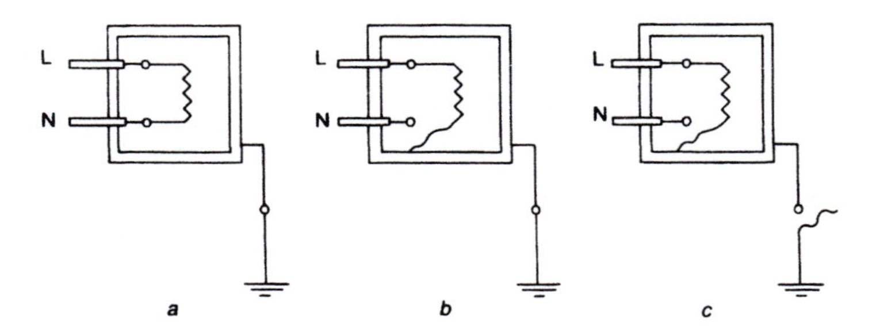

Figure 3.6 Effect of the loss of an undetected earth return circuit

a Normal condition
b First undetected fault causes load current to return through protective earthing conductor
c Second fault (loss of earth return) leaves the housing (exposed-conducting-part) live at full mains voltage to earth

not necessarily hold up when an earth fault occurs. The fault current may then be insufficient to operate the protective device although the voltage might still be at a dangerous level.

3.6 Reducing the risks

Additional safety can be provided by the inclusion of an earth leakage detector which monitors the balance between the load current and the return current. Such detectors, in the form of a miniature core balance relay, were developed in the early 1960s and are being increasingly used throughout the world. The principle of these so-called residual current devices or RCDs* is shown in Figure 3.7. When no earth leakage currents are present, I_L and I_N are equal and opposite and no magnetic flux is induced in the ring transformer. In the case of 3-phase circuits, it will be the vector sum of the phase currents which will be equal and opposite to the resulting current returning through the neutral.

If there is an earth leakage current I_E escaping *beyond* the RCD, the sum of the currents through the toroid will have a resultant value of I_E which will create a flux in the iron or ferrite ring and so induce a voltage in a detector coil wound on the toroid. This signal—amplified if necessary—can be arranged to give an alarm indication or to trip the power circuit at a preset value of I_E. An RCD will also detect leakage currents from the neutral to earth.

Differences between line and neutral currents of less than 0·03% can be reliably sensed, so that circuits with this type of protection can be at least 10 000 times

*These devices were previously known as current-operated earth-leakage circuit-breakers, or ELCBs in UK commerce and as ground fault circuit interrupters, or GFCIs in American commerce. In Central Europe they are known as FI switches. The IEC has adopted the somewhat esoteric term 'residual current device', or RCD which has been incorporated into the 15th and 16th editions of the IEE Wiring Regulations. The term RCD is accordingly used here.

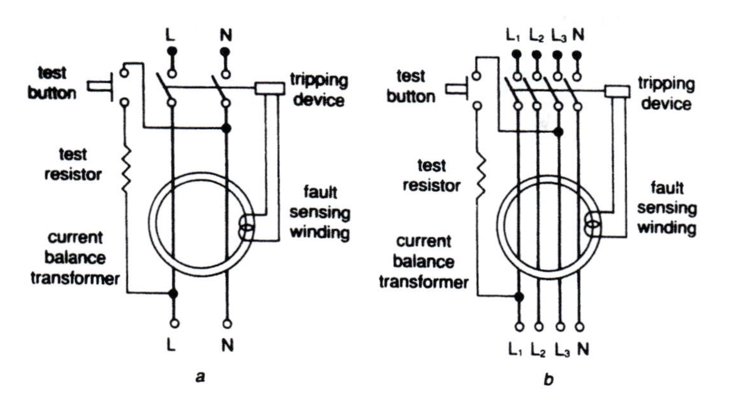

Figure 3.7 Diagrams showing the principle of the residual current circuit breaker device (RCD) to detect fault and leakage currents to earth and interrupt the supply

a Single-phase unit
b Three-phase unit

more sensitive to earth faults and earth leakage than a simple fused circuit, and some 5000 times more sensitive than an overload relay set to trip at 50% overload.

These devices are designed with an integral circuit-breaker giving overload, short-circuit and earth-leakage protection and having standard modular widths, making them interchangeable with miniature circuit-breakers which do not have earth-leakage protection. A push-button on the case enables the rated earth-leakage sensitivity to be tested by connecting a suitable high resistance across line and neutral. Because they can sense a few milliamps of earth leakage, they do not require a low earth loop impedance in order to function and can, in fact, protect from lethal shock even if the installation has neither earthing nor bonding of any exposed- or extraneous-conducting-parts.

Thus it will be seen that, for the standard low voltage (400/230 V) installations with earthed neutrals, the use of current balance protection overcomes all of the dangerous conditions with regard to fire and electrocution referred to in Section 3.5.

Typical RCD units are designed for load currents of 63 and 100 A with various earth-leakage sensitivities such as 5 mA, 30 mA, 100 mA etc. To obtain discrimination between main and sub-circuits, RCDs can be supplied with a built-in time delay. For 'instantaneous' units, standards have stipulated that they should not trip at less than half the rated sensitivity. At the rated leakage current they should trip within 0·2 seconds and at 5 times the rated earth leakage the circuit-breaker must isolate the power supply within 30 ms.

It should be noted that an RCD does not prevent electric shock which can of course be frightening and unpleasant, but it will respond rapidly to current through the body and will trip the supply before fibrillation has time to set in, i.e. within one heart beat. Note also that the current through the body is determined not by the RCD sensitivity but by the voltage and circuit resistance—mainly body resistance. Assuming this is 1000 ohms and neglecting external loop resistance, a 230 volt shock will subject one to 230 mA through the body. It is generally accepted that an RCD rated at 30 mA will provide a very high degree of protection against ventricular fibrillation and hence against death. A more sensitive rating would not reduce the shock current although it might trip in a slightly shorter time.

The IEE Wiring Regulations specify the use of 30 mA RCDs in certain instances to give enhanced safety. The 15th edition required these to trip in 40 milliseconds or less with a leakage current of 250 mA, whereas the 16th edition stipulates that the device should trip within 40 ms at a leakage of 150 mA.

The suitability of the 30 mA rating was argued in a German report which stated that of 576 known fatal accidents due to electricity, 539 could have been prevented by the use of a 30 mA RCD. In the remaining 37 cases the victims were electrocuted by a phase to phase or a phase to neutral current involving no earth-leakage signal.

Higher sensitivities are possible but there is then an increased risk of spurious tripping. This is mainly due to capacitance currents in the circuits downstream of the RCD. Even though the capacitance to earth may be balanced as between line and neutral (i.e. 'common mode' capacitance) it is only the line which will have a substantial voltage to earth and this will produce some unbalance and hence residual flux in the RCD. The connection of equipment having interference suppression capacitors between line and earth can increase the unbalance to a level sufficient to cause unnecessary tripping.

Capacitance to earth will of course have a lower impedance with respect to high frequencies. Because the release mechanisms of modern RCDs can respond to flux pulses as short as a millisecond and, because sharp voltage spikes contain high frequency components, sensitive RCDs are also liable to be tripped by switching transients. In these cases the harmonic (high frequency) components are shunted to earth through the overall system capacitance and the RCD reads this as a leakage current—which it is. Instead of reducing the sensitivity of the device, stability could no doubt be improved by introducing some inductance in front of the RCD to suppress the higher harmonics in the voltage spikes.

False tripping may also occur where inductive circuits containing relay coils etc. are interrupted beyond the RCD, producing a steep transient voltage rise. Problems of this type can be solved by connecting a suitable capacitor across the inductive load, or by using a discharge circuit across the inductance.

It has become common practice in new installations to include RCD protection on some of the circuits in general purpose house service units, as can be seen in Figure 3.8.

Although in the UK 30 mA is the most sensitive rating commonly used, in North America 5 mA has been adopted for shock protection because experiments had shown that the let-go level for 99·5% of the population was 9 mA for men and 6 mA for women. In other words, above 5 mA it was considered necessary for the RCD to do the 'letting go'. Many millions of these units are now installed in the USA and Canada.

High sensitivity RCDs have also been used extensively in France since the 1960s and, after several years experience, EdF stated that it knew of no instances of personal accident following the touching of live or defective equipment in an installation protected by a high-sensitivity RCD.

The adoption of RCDs in the UK has been somewhat slower, apparently because of the leakage currents which flow in electrical apparatus accepted in the British market, leading to a fear of false tripping at 30 mA. Improvements in product

Figure 3.8 *House service unit equipped with miniature circuit-breakers (MCBs) and with some circuits protected by a residual current device (RCD)*

Courtesy MK Electric Limited, London

insulation and in the design and sophistication of the current balance earth-leakage detector will undoubtedly lead to their wider use.

Metal-clad, hand-held equipment and all appliances used outdoors should only be connected via an instantaneous 30 mA RCD. Plug-in units, as shown in Figure 3.9, are in common use in the UK for individual socket outlets.

3.7 Electricity supply systems—principles and practice

Until the standardisation of 'electricity' in the UK there were many different supply and distribution systems: various voltages at the meter, 2-wire and 3-wire d.c., single-phase, 2-phase and 3-phase a.c. and several different frequencies.

The 1926 Electricity Supply Act empowered a Central Electricity Board (the CEB) to standardise the frequency of all supplies at 50 Hz with a single-phase voltage of 230 to the domestic consumer. This was then raised to 240 V but has now reverted to 230 since agreement has been reached in the EC to adopt a common European voltage of 400/230 V 50 Hz (mainland Europe previously used 380/220 V 50 Hz).

The amended Electricity Supply Regulations (1994) require the supplied voltage to vary by not more than +10 –6% from the nominal 230 V at the supply terminals, but the permissible minimum voltage regulation is not specified (voltage regulation

Figure 3.9 Typical plug-in type RCD to protect portable and outdoor apparatus — 13 A rating, 30 mA sensitivity

in this context means the loss of terminal voltage between no load and full load due to the internal impedance of the source of energy). The minimum regulation is important to the consumer since it determines the prospective short-circuit current (PSCC). For example, with a supply having 5% regulation it is necessary to install incoming switchgear or fuses able to clear a PSCC of 20 times full load current, whereas with a 0·1% regulation the PSCC would be 1000 times full load current.

Because the PSCC is a critical factor in circuit design, the Electricity Supply Regulations stipulate that the supplier shall state the PSCC at the consumer's terminals. In fact, electricity undertakings are often not able or not prepared to do this because the voltage regulation and hence the PSCC is affected by variables such as the number and size of sub-station transformers connected into the network at any given time.

In practice a 'pessimistic' upper limit of 16 000 amps is generally assumed for the PSCC for domestic supplies. In rural districts it is more likely to be in the region of only 1000 amps.

In general, short-circuit faults present few problems apart from loss of supply and it is earth faults and earth leakage which require more careful consideration. During an earth fault the resistance of the return circuit through the earth loop to the transformer neutral will usually have a far higher resistance than that of the line. At the fault the earthed conducting part will thus be at nearly line voltage above any other earthed parts in which no current is flowing—hence the need to bond 'extraneous conducting parts' to 'exposed conducting parts'.

This voltage elevation in close proximity to earth faults presents a hazard which is a complex aspect of electrical safety. The subject is explored in some detail in the following section.

3.8 Characteristics of supply

The 230 V single-phase service to domestic consumers and others is normally provided via a dedicated 'service line' which is tapped off from a local 'distributing main'. These will be overhead or underground depending on the locality. In addition to the line and neutral, the supply undertaking will in most cases install an earthing terminal giving a continuous metallic path to the earthed star point of the distribution transformer.

Underground cable systems have in the past used cables with a separate earth conductor in the form of a metal sheath or armouring. Alternatively a separate earth core may be included in the cable. When the neutral and earth conductors are separate in this way, it is termed a TN-S system by the IEE Wiring Regulations which are based on the terminology of the International Electrotechnical Commission (IEC). The T stands for terre (earth), N for neutral and S for separate. The Electricity Supply Industry (ESI) still tends to refer to it as an SNE system (separate neutral and earth). The TN-S type of supply is shown diagrammatically in Figure 3.10*a*.

The 1988 Electricity Supply Regulations permit the earth conductor and the neutral conductor to be combined under certain conditions, in order to provide a low impedance earth loop. It is then termed a TN-C (or CNE) system. The combined neutral and earth conductor may be one of the cable cores or may be the sheath or armouring. A single-phase cable of the latter type has only one core conductor and is accordingly termed a concentric cable.

Five methods of providing an earth fault loop are shown diagramatically in Figure 3.10.

Before proceeding further it will be helpful to list some special definitions which have been adopted on the subject of 'earthing'.

Circuit protective conductor (CPC) (more commonly known as the earthing conductor)	*The connection between exposed-conducting-parts and the main earthing terminal of an installation*
Earthing conductor	*The connection between the main earthing terminal of an installation and the earth electrode or other means of earthing*
Potential equalisation conductor* (PE) also; Supplementary bonding conductor, Equalisation bonding conductor, Main bonding conductor	*Connections between exposed-conducting-parts and other metalwork (extraneous-conducting-parts) to keep them substantially at the same potential under fault conditions*
Protective conductor#	*Any of the above conductors*
PEN conductor	*A combined neutral and protective conductor*

*In many European countries the earthing terminal on electrical apparatus is known as the potential equalisation terminal and is referred to on drawings etc. as PE.

#Previously known as earth continuity conductor.

Note: In rooms containing a fixed bath or shower, it is considered necessary to bond between simultaneously accessible extraneous-conducting-parts, even when there is no electrical equipment present. This is then termed supplementary equipotential bonding.

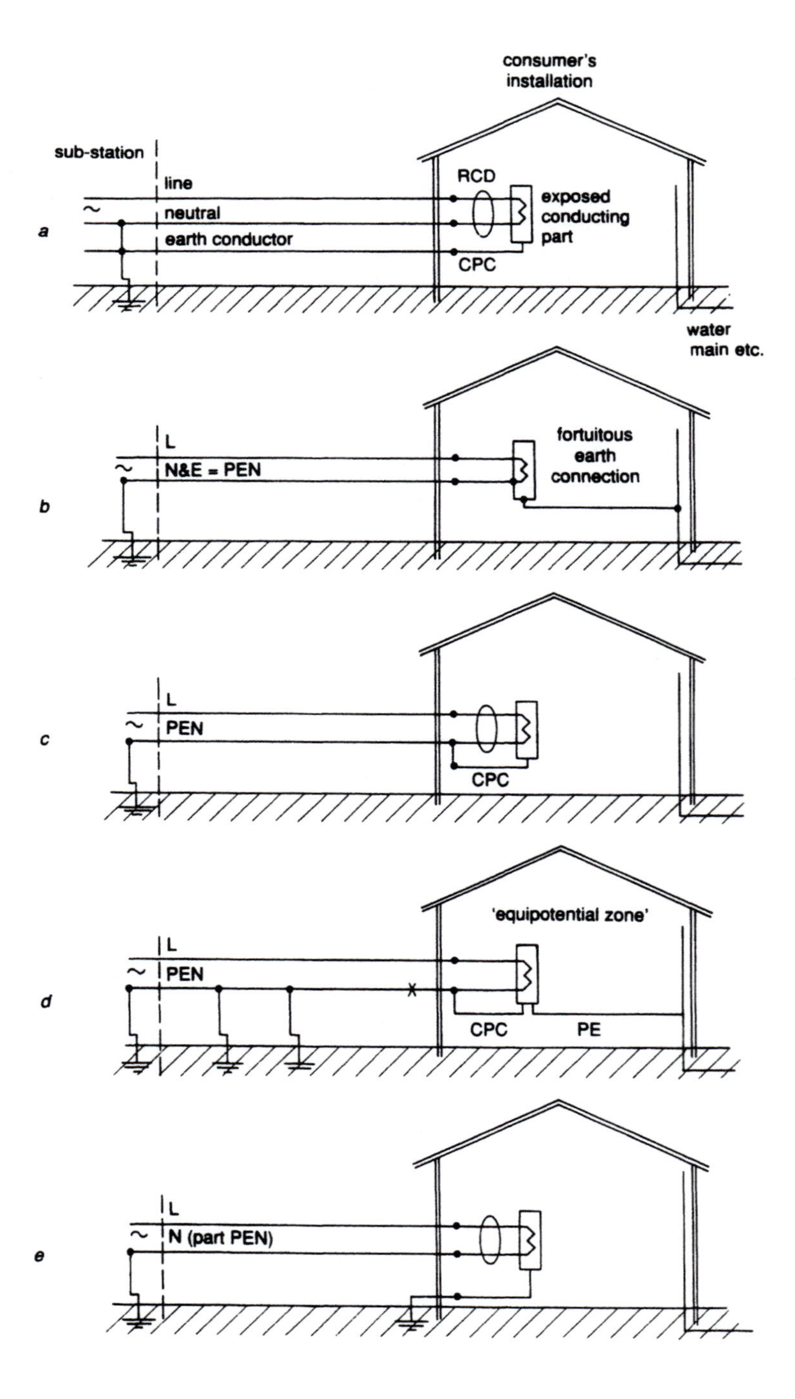
consumer's installation
sub-station
line
neutral
earth conductor
RCD
exposed conducting part
CPC
a
water main etc.
L
N&E = PEN
fortuitous earth connection
b
L
PEN
CPC
c
'equipotential zone'
L
PEN
CPC
PE
d
L
N (part PEN)
e

Systems incorporating combined neutral and earth conductors can be divided into 3 types, described below under *b*, *c* and *d*. A high resistance earth loop in conjunction with an RCD is described under *e*.

(*b*) No earthing terminal is provided and the consumer connects all exposed-conducting-parts directly to the neutral (PEN). Special authorisation is required since load currents are enabled to flow through the exposed-conducting-parts. As can be seen from Figure 3.10, RCD protection is not possible in this case.

(*c*) Separate neutral and earthing terminals are provided, these being connected together outside the installation. In other words the combined neutral and earth are split before they reach the consumer's installation. The supply system is thus TN-C and the installation is TN-S. The system as a whole is termed TN-C-S. As can be seen from Figure 3.10, this enables the consumer to use RCD protection.

(*d*) Before reliable RCDs became available, the problems of providing and maintaining a reliable low impedance earth fault return path for consumers supplied from an overhead line or from a cable system without a continuous earth conductor led to the increasing use of a system known as protective multiple earthing (PME). This differs from the basic TN-C or TN-C-S systems in a number of respects, the major one being that the resistance between the PEN conductor and the ground may be as high as 20 ohm. The earth fault loop impedance, however, will be less than this (normally less than 1 ohm) since the PEN conductor is used as the earth fault return path in addition to its normal load carrying function as a neutral.

The main hazard with PME is that if the neutral (PEN) conductor becomes disconnected or acquires a high resistance at the consumer's earthing terminal, it can leave the earth terminal standing at line potential via the consumer's electrical equipment. Exposed-conducting-parts connected to the earthing terminal will accordingly also then be at line potential (see Figure 3.10).

Obviously, the voltage elevation of a casing (exposed-conducting-part) due to a broken or high resistance PEN circuit may not be accompanied by a similar voltage elevation of the ground in the vicinity and, as previously described,

Figure 3.10 Alternative earth loop systems for single phase low voltage supplies

a TN-S (or SNE) system: Earth fault is completed via cable sheath and/or earth core in the cable
b TN-C (or CNE) system: N and PE conductors are the same, e.g. where the sheath of a buried cable is the return conductor for both load current and earth fault currents. Protection by RCD is not effective in this case
c TN-C-S system: N and PE conductors are the same as far as the consumer's terminals and are then divided. The RCD is effective because leakage and fault currents to earth do not pass through the toroid of the RCD
d Authorised TN-C-S supply known as PME (protective multiple earthing): Any loss of continuity in the PEN conductor (at **X** for instance) could elevate the potential of exposed-conducting-parts with respect to extraneous-conducting-parts. Equipotential bonding is therefore required
e TT system, i.e. separate earthing: A separate earth electrode is connected at the consumer's installation. No earth terminal is provided by the supply undertaking so that an earth-leakage sensing device is essential due to possible high earth loop impedance

this can lead to a dangerous potential difference between exposed-conducting-parts and extraneous-conducting-parts. It has therefore been considered essential with PME to provide equipotential bonding throughout the installation. That is to say, an installation supplied from a PME system must—according to the 1988 Electricity Supply Regulations—be within what is termed an equipotential zone. Furthermore, the Regulations specify bonding conductors which are proportional to the cross-section of the PEN conductor, as follows:

PEN mm^2	PE mm^2
35 or less	10
>35 up to 50	16
>50 up to 95	25
>95 up to 150	35
>150	50

(*e*) If the supply undertaking is unable to provide an earthing terminal of a suitably low impedance, the consumer will need to find his own independent 'earth'—normally by means of a local earth electrode. In some areas all newly constructed property must be equipped with its own earthing electrode.

Where the consumer has a separate earthing arrangement from the sub-station earthing system it is known as a TT installation. As such it may have a very high earth return impedance through the ground and must accordingly be protected by RCDs, which as we have noted, can function with high earth loop impedances. A 30 mA RCD for example will operate with an earth loop impedance Z_s of up to

$$Z_s = \frac{230}{0 \cdot 03} = 7667 \text{ ohm}$$

With very high loop impedances low values of earth current, which are insufficient to trip the RCD, can cause exposed conducting parts to remain at a dangerous potential. Taking 50 V as the maximum allowable safe value, it is necessary to ensure that the RCD will trip before the product of the earth leakage current and the earth loop impedance Z_s exceeds 50 volts, i.e. the RCD rated sensitivity $I_{\Delta n}$ must be such that

$$I_{\Delta n} \times Z_s < 50.$$

If this is ensured equipotential bonding may not be necessary. A TT system may also be used where it is not possible to create an equipotential zone.

Thus it will be seen that the PME system relies, for safety under fault conditions, on equipotential bonding. The more recent TT system depends upon sensitive RCD protection but requires some determination of the earth fault loop impedance in order to specify the RCD sensitivity. Methods of measuring the loop impedance and electrode resistance are accordingly addressed in the following section.

Meanwhile we may note that the maximum permissible earth loop impedance Z_s for a TT system protected by a 30 mA RCD will be given by the equation

$$0 \cdot 030 \times Z_s = 50$$

so that

$$Z_s = 5000/3 \text{ or } 1670 \text{ ohm.}$$

3.9 Earthing systems

The connection of an electrical power system to earth is normally by means of one or more conducting rods driven two or three metres into the ground. The resistance of these so-called earth electrodes comprises the contact resistance of the rod with the ground or soil and the resistance of the ground itself, these two values being effectively in series.

The overall resistance will be very dependent on the nature of the ground and its moisture content. The electrodes will each usually have a resistance to earth of a few tens of ohms. This can be regarded as a contact resistance. The resistance of the soil however, may range from some two or three hundred ohm-cm in marshy ground, to nearly a million ohm-cm in dry sand.

It may be noted here that ground resistivity, in ohm-cm, is equal to the resistance across the faces of a centimetre cube, but in the vicinity of an earth electrode, the actual resistance will not be proportional to distance because of the decreasing current density.

Where the conductivity of the ground is low, it is difficult and sometimes costly to obtain either a low or a consistent earth resistance. Cable sheaths cannot always be relied on to give a continuous earth return, particularly with old buried cables—perhaps used in the past for obsolete direct current systems. Such problems of high earth loop impedance have been overcome to some degree by the use of PME, i.e. the protective multiple earthing of a PEN conductor.

PME systems need to be designed so that all parts of the main distribution cable neutral have two separate routes to an earth electrode. In addition the resistance to earth of this neutral (PEN) conductor must not exceed a safe value (20 ohms in the UK). Although supply undertakings usually provide a main earthing terminal, this is not obligatory and without an earthing point for his installation the consumer has to find a suitable earth loop circuit. In the past this was commonly obtained by connecting to the domestic water main. Nowadays, the widespread use of plastic service pipes has led to the practice of having an installation electrode driven into the ground adjacent to the property.

This forms a TT installation and the earth loop impedance will then consist of the following impedances in series, as shown in Figure 3.11.

(a) Transformer winding
(b) Phase supply lines to the consumer
(c) Earth fault on consumer's premises
(d) Consumer's PE conductor
(e) Consumer's electrode resistance
(f) Soil resistance to PEN conductor
(g) PEN conductor to star point of transformer (normally $\leq 0 \cdot 35$ ohm)

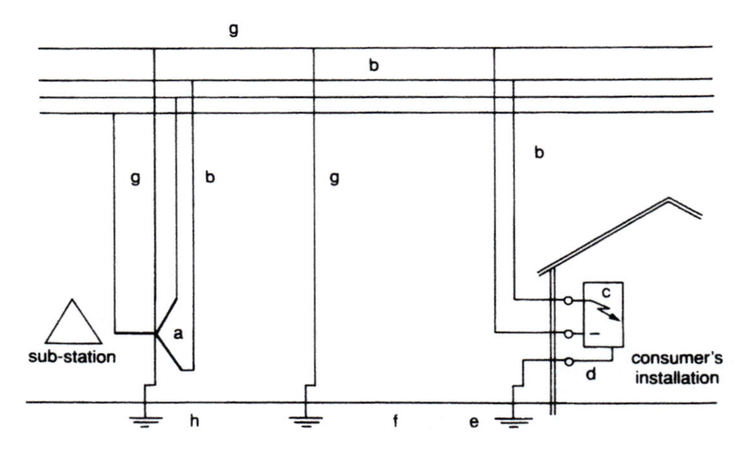

Figure 3.11 *Components of an earth fault loop circuit for a single phase TT installation*

a Effective impedance of sub-station transformer winding
b Phase supply conductors to the consumer
c Earth fault impedance
d Connection from exposed conducting part to consumer's earth electrode
e Resistance to earth of consumer's electrode
f Soil resistance to PEN conductor
g PEN conductor to star point of sub-station transformer
h Resistance of parallel circuit through the ground to transformer star point

In parallel with (f)-(g) will be:

(h) The resistance through the ground to the sub-station electrodes.

Only (e) and (f) are likely to be significant and it is thus necessary to ensure that, for a TT system, the consumer's electrode and soil resistances are within acceptable limits in relation to the sensitivity of his earth fault protection.

3.10 Measurement of electrode contact resistance and ground resistance

The resistance between an electrode and the ground can be measured as indicated in Figure 3.12. It is necessary for the voltmeter impedance to be considerably greater than the resistance of the instrument electrode Y so that the volt drop at Y will be negligible compared with the volt drop at X.

The electrode contact resistance to the ground will then be given by the voltmeter reading divided by the ammeter reading, $R = E/I$. Several measurements may be required with varying electrode spacings in order to obtain repeatable results. Wider spacings will be needed for the determination of low values of electrode resistance and, if the contact resistance is 1 ohm or less, distances of several hundred metres may be required. This is because the electrode contact resistance includes the ground surrounding the electrode and the radius of this effective electrode resistance area is an indeterminate factor influenced by the ground resistivity.

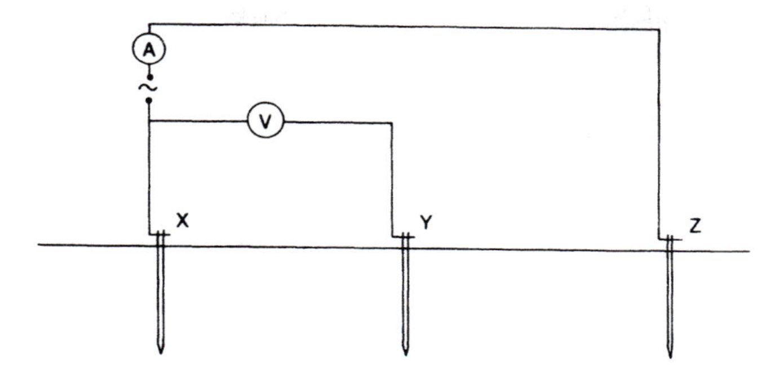

Figure 3.12 *Measurement of resistance to ground of earthing electrode (X)*

If resistance of Y is negligible compared with the resistance of the voltmeter,

$$R_X \text{ is given by } \frac{\text{voltmeter reading}}{\text{ammeter reading}}$$

Electrode Z needs to be at least 30 m from X and Y about midway between. Wider spacings may be necessary to avoid overlap of the electrode effective resistance areas, particularly if X has a low resistance value

The soil or ground resistivity itself can be similarly measured using four test electrodes as indicated in Figure 3.13. Again it is important to ensure that the resistance areas of the electrodes do not overlap.

In both the above cases, serious errors and wandering of the voltmeter needle are liable to occur due to earth currents. To overcome this, a non-standard frequency may be advisable for the test currents.

Earth currents are not a recent phenomenon and are not only due to man-made electricity. In 1902 for example, there was a discussion at the Institution of Electrical Engineers in London on the subject of earth currents derived from

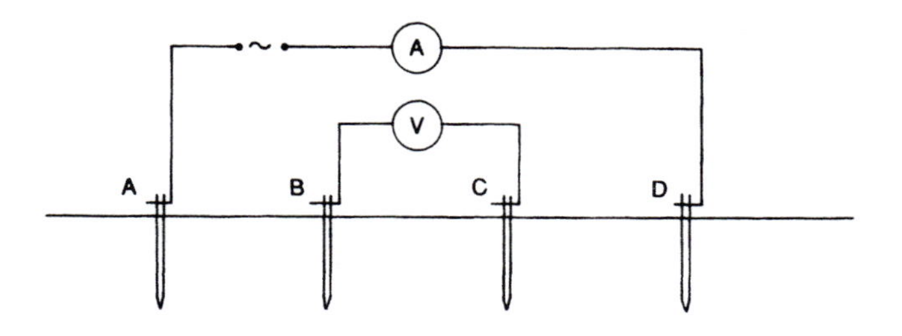

Figure 3.13 *Measurement of earth resistance*

Soil resistance between B & C is given by

$$R_{b-c} = \frac{\text{voltmeter reading}}{\text{ammeter reading}}$$

As for the electrode resistance measurement, it is important to ensure that the effective resistance areas of the electrodes do not overlap. Resistivity ρ will then be proportional to the value of R_{b-c}

distribution systems, which included the following contribution from W.M. Mordey:

> 'This Institution started as the Society of Telegraph Engineers. In the early days earth currents were considered a good deal on account of their bearing on telegraph working, as may be seen by the records of some of our early meetings—there are eighteen papers on the subject in the first ten volumes of our Proceedings. Telegraph engineers need not be reminded that all the old needle instruments were made with rotary dials, so that, when earth currents came on, the dials could be turned round to bring the needle to the middle position whatever the force or direction of the earth current might be. I remember, when I was in the telegraph service, measuring earth potentials on a line seventeen miles long, and finding forty Daniell cells necessary to counteract the earth current. That meant a potential of nearly two volts per mile. I remember at the same time I ran a little electric motor by those earth currents on the telegraph circuit. That was long before tramways or electric light. I suppose nowadays we should be told that that motor was being run by leakage currents from tramways, . . .'

The measurement of electrode and soil resistance is seldom necessary since the sole purpose is to determine the overall earth loop impedance which can usually be more directly ascertained. An approximate and simple way to check that the earth loop impedance is within required limits is to connect a suitable load, such as an electric kettle, across line and earth and to measure the loss of voltage when the load is switched on.

For example, if a no-load voltage of 230 V falls to 226 V when a 1·5 kW (= 6·52 A) load is connected, the earth loop impedance will be given by

$$Z_s = \frac{230 - 226}{6{\cdot}52} = 0{\cdot}61 \text{ ohm.}$$

In carrying out this test care must be taken to ensure that no danger would arise if the earth circuit is defective or incomplete. It is also necessary to disarm any RCD in the circuit, as one is effectively applying an earth fault.

3.11 Review

The predominant system of low voltage electrical supply uses an earthed power source with various methods of protection against electric shock and incendiary conditions. To cater for all likely and foreseeable faults and modes of failure, very complex rules and regulations have been developed and, in attempting to be concise, official documents on the subject are not always readily understandable. The 1991 edition of the IEE Wiring Regulations, for example, continues to generate explanatory publications and many of the regional electricity supply companies have issued their own guidance booklets on protective multiple earthing. However, although PME is coming into increasing and more general use, it has yet to be formally defined.

The legal requirements in the UK for PME are given in the Electricity Supply

Regulations 1988 but these are so disgracefully obscure as to be an insult to the reader. Section 7 for instance entitled 'Protective multiple earthing' includes the following Rule:

(3) (a) This paragraph applies only where—
 (i) at least one of the consumer's installations (not exceeding four in total) whose connections to a distributing main lie nearest to the end of the main uses the supply neutral conductor for the purpose of connecting the installation with earth; and
 (ii) the distance of the furthest of those connections from the end of the distributing main does not exceed 40 metres.
(b) In any case where this paragraph applies the supply neutral conductor shall be connected with earth at a point no nearer to the source of voltage than the junction between the distributing main and the service line connecting the consumer's installation referred to in sub-paragraph (a) above which is nearest to the source of voltage in the distributing main.

To describe the problems and solutions, for the provision of safe LV systems, in a reasonably transparent style has required a fairly detailed and somewhat superficial treatment in which a number of particular aspects have been omitted. These include the earthing of high voltage supplies and of HV/LV sub-stations and the provision of 'clean' earths for electronic circuits.

3.12 Bibliography

1 BS EN 60529: 1991 Specification for degrees of protection provided by enclosures (IP code)
2 IEE Wiring Regulations, 16th Edn., 1991 (= BS 7671 : 1992)
3 AMOS, S.W.: 'Dictionary of Electronics' (Butterworths, 1986)
4 KIDD, A.L.: 'Electrical protection for domestic and small commercial installations', *PEJ*, Jan. 1987
5 NOWAK, K.: 'Fifteen years of current operated earth leakage protection in the VDE regulations', *Der Elektromeister & Deutches Elektrohandwerk*, Dec. 1973, **48**, (24), pp. 1771–1774
6 'FOCUS ON ELCBs', *Electrical Times*, June 1976, pp. 10, 11, 13
7 PICKEN, D.A.: 'Electrical installation practice in some Western-European countries', *Proc. IEE*, July 1967, **114**, (7)
8 ENGINEERING RECOMMENDATION G.12/2, Code of Practice for Protective Multiple Earthing (Electricity Council, 1982)
9 EASTERN ELECTRICITY Notes of Guidance on Installations to be connected to Protective Multiple Earthed Systems, March 1991
10 THE ELECTRICITY SUPPLY REGULATIONS 1988 (HMSO)
11 THE ELECTRICITY SUPPLY (AMENDMENT) REGULATIONS 1990 (HMSO)
12 FINLAY, G.S.: 'Earthing facilities provided by electricity companies', ERA Conference 26–27 Feb. 1991, *ERA Report 90-0701*
13 CP 1013: 1963, Earthing [BSI]
14 BS 7430: 1991, Code of Practice for Earthing
15 WEDMORE, E.B.: 'Earth currents derived from distributing systems', *J. IEE*, 1901–1902, **XXXI**, pp. 576–610

Chapter 4

Cables and fires

The title of this chapter is not meant to imply that cables are a major cause of fires, but rather that the organic materials used to insulate and protect the conductors contribute to the severity of fires in buildings, tunnels, ships etc. In the United Kingdom the classic cable types have been paper-insulated lead-covered single-wire armoured (PILCSWA) cables for underground power distribution and vulcanised india rubber (VIR—or vulcanised rubber insulated, VRI) wiring for internal distribution. The sulphur used in vulcanising the rubber attacks copper so that it is necessary for rubber-insulated copper conductors to be tinned.

Cables within buildings have been protected usually by a lead sheath over the insulation. For certain duties an all-rubber cable was used with an outer rubber covering, these cables being known as tough rubber sheath or TRS cables. When laid up to be semi-flexible with a circular section they were, and often still are, known as cab tyre sheath or CTS cables. Although paper insulated cables are still in use in this country and to a considerable extent elsewhere - their life in undisturbed ground being virtually unlimited within their rating - natural rubber cables have been progressively replaced by synthetic formulations. These include the elastomeric and thermoplastic materials shown in Table 4.1. An elastomer is a rubber-like substance with some elasticity; a thermoplastic substance deforms permanently and may melt when heated. All these cable insulating materials can be assessed and compared by a large number of different mechanical and electrical properties. From the point of view of fire behaviour, the most important are

- Oxygen index: the minimum percentage of oxygen in a mixture of oxygen and nitrogen which will sustain combustion. If this is less than the proportion of oxygen in the atmosphere (21%), the material will burn freely in air.
- Self-ignition temperature: the temperature at which it will start to ignite.
- HCl emission: the percentage by weight of hydrochloric acid gas emitted during combustion.
- Smoke emission: the density of the smoke emitted after specified times of subjection to a standardised flame in a standardised test cell.

At one time the most important aspects of cable safety characteristics were considered to be the insulation resistance and mechanical strength, or ruggedness. Over the years it has been realised that cable fires are rarely caused by failure

Table 4.1 Some insulating materials commonly used in power cable construction

Elastomeric	
natural rubber	NR
silicone rubber	SR
ethylene propylene rubber	EPR
ethylene vinyl acetate	EVA
cross-linked polyethylene	XLPE
polychloroprene	PCP
chlorosulphonated polyethylene	CSP
Thermoplastic	
polyvinyl chloride	PVC
polyethylene	PE
polypropylene	PP
Other	
paper insulation	PI
mineral insulation	MI
polytetrafluorethylene	PTFE

of undisturbed cables themselves, which are almost invariably protected by fuses and circuit-breakers. However, in many industrial premises the quantity of organic combustible material in busy cable runs is nowadays a major contribution to the total fire audit. In some modern plants, such as in power stations, the cable runs may actually be the only large bank of combustible material. The safety aspects of cable installations have accordingly led to an increasing need for cables having minimal emissions of smoke, toxic fumes and acid gas under fire conditions.

At the same time the essential requirement for an electric cable is that it will be free from electrical faults over its working life, with a reasonable factor of safety. That is to say it must be capable of withstanding its rated voltage and of passing its rated current without distress. The rated voltage can easily be confirmed by an over-voltage or flash test. It is normal practice to apply a quality control test before installation of twice the rated voltage plus 1000 V for LV cables and to apply a test voltage of 80% of this after installation. Various voltages are used to test HV cables, depending on the rated voltage and the material.

The current rating is a function of the maximum safe working temperature and this can only be assessed in a test laboratory.

One method of determination of the maximum permissible working temperature is by the extrapolation of short term tests at elevated temperatures. It relies on the principle that the rate of deterioration of insulation under the action of heat increases with temperature, i.e. the higher the temperature, the shorter the working life. It enables an extrapolation to be made down to a temperature low enough to give an acceptable length of working life. A more detailed description of the method is given at the end of this chapter.

Cables are normally rated for standard system voltages, such as 600 and 1000 V for LV cables. The current ratings will depend on the thermal conditions in which the cable is used, for instance,

- Maximum ambient temperature
- Number of cables bunched together
- Characteristics of the overload protective device
- Whether the cable is run in the ground, on a rack, in plaster etc.

It will be clear that, with so many variables, tables of current ratings become very complex. Added to which, for practical purposes load diversity needs to be taken into account when sizing distribution cables.

Characteristics of some commonly used types of cable are considered in the following sections.

4.1 Polyvinyl chloride cables and wiring

Polyvinyl chloride (PVC) was developed in Germany in the 1930s as an alternative to rubber. It is a thermoplastic resin formed by the polymerisation of vinyl chloride (CH_2:CHCl) and is resistant to moisture, dilute acids and alkalis. The monomer is carcinogenic. In addition to the merit of being less flammable than rubber, PVC also enables untinned copper conductors to be used. It has the demerit of a relatively high dielectric loss, making it of doubtful suitability for a.c. voltages over 6·6 kV. At higher voltages suitable mechanical protection and sensitive overload and short-circuit protection are necessary. PVC cables have been used in Germany up to 20 kV, but improved alternative materials are now available.

Being a mixture of PVC resin, plasticiser, stabiliser and extender, a wide range of properties is possible. As much as 50% plasticiser can be added to the brittle base resin. Stabilisers are added to combat evolution of hydrochloric acid gas at elevated temperatures. At 175°C two per cent may be given off in an hour and even at 120°C the acid gas may be slowly liberated. Overloads are thus to be avoided with PVC cables because of this chemical change, as well as their thermoplastic nature.

Polyvinyl chloride insulation is, in its standard formulation, designed for a maximum working temperature of 70°C. This means that the current-carrying capacity of PVC wiring and cabling decreases rapidly with increasing ambient temperature and falls to a current rating of zero at an ambient temperature of 70°C. By the same token, PVC cables running under thermal insulation alongside central heating pipes—in a roof space for example, or in a closed floor duct with steam pipes—are liable to overheat. To some extent this will be self-correcting since the softening of the insulation at high temperatures will be reduced as the plasticiser volatilises away. The removal of plasticiser is also found to be accelerated by the leeching effect of many types of thermal lagging. PVC affected in this way reverts to the natural brittleness of the pure resin and is therefore liable to crack if disturbed.

Short-circuit tests have shown that a conductor peak temperature limit of 130°C should be imposed on armoured, PVC-insulated cables where they are likely to be subjected to repeated short-circuits. A maximum conductor temperature of 160°C may be used for a single short-circuit. One of the weaknesses of this material is its tendency to creep. The single strand conductors used in the cores of low voltage wiring impose a stress on the insulation if the cable or wiring is bent. The softened insulation becomes progressively thinner at the outside of the bend and eventually splits and relaxes across the corner, leaving the conductor out at the elbow. Besides softening and being liable to flow under stress when warm, PVC becomes increasingly brittle as the temperature is lowered. For this reason, although PVC wiring can be used at low temperatures, it should not be installed or drawn into conduit when the ambient is below 0°C. In an emergency it can be stored in a warm room and put in place before it gets cold.

There are several grades of PVC the properties of which are largely governed by the quantity and type of plasticiser used, generally castor oil. The polymer on its own is a horn-like inflexible substance and, when the plasticiser leeches out or migrates, the cold-crack temperature of the compound will rise. PVC cable suspended out of doors will weather and suffer hardening due to loss of plasticiser. After a few years, movement due to wind in a cold season may then cause cracking.

Hard grades of PVC, containing less plasticiser, are not so subject to flow when heated for a short period such as while soldering. This benefit, however, is at the expense of increasing brittleness at lower temperatures.

High temperature grades of PVC incorporate a non-vaporising plasticiser so that the compound is able to retain flexibility after long periods at temperatures in the region of 100°C. These grades are intended for connecting to terminal boxes in the vicinity of lighting fittings and domestic immersion heaters. Sustained temperatures above 100°C are not possible with PVC cables as the polymer itself will then have a limited life due to decomposition.

Power cables insulated with PVC are normally wire-armoured, but no metal sheath is necessary since moisture has very little effect on the insulation.

If a plastic mat is placed under a porous flower pot on a French-polished table, the surface of the table beneath the mat will turn white. This shows that, although a plastic sheet is effectively water-proof, it is by no means vapour proof. It is perhaps not always realised that, with any plastic cable material—no matter how thick—ionic diffusion will ensure that the moisture content at the core will in due course reach a value equivalent to the long term average humidity outside the cable. The insulation must therefore work satisfactorily with this equilibrium moisture content. Metal barriers will prevent diffusion only if they provide a hermetic seal. Moisture, as distinct from vapour, absorbed by degraded or weathered insulation will decrease its electrical resistance and may cause overheating and breakdown if a high voltage is applied with insufficient time for the moisture to be driven off.

4.1.1 Fire and smoke

Laboratory tests on single PVC cables show that when a flame is applied and then removed, the cable is self-extinguishing. That is to say, the combustion of the insulation does not persist. For many years it was assumed that PVC insulation was effectively fire-proof in that it did not add substantially to an overall fire load in a building. It is true to some extent because, after a house fire where there are both the older rubber-insulated cables and PVC cables, the rubber cables but not the PVC cables are found to have carried the fire from one area to another.

In a fire at a power station in Italy in 1967 a major conflagration was propagated along both horizontal and vertical PVC cable runs. This and a number of other large fires involving PVC showed that if the total mass of insulation was above a certain concentration it was not self-extinguishing. It applies particularly in vertical shafts and in cable tunnels where convection currents can fan the flames. As an example of this, a fire in a cable tunnel in the USA destroyed over 100 km of PVC cable and liberated 15 tonne of hydrogen chloride gas, with vast associated losses. Subsequent tests have shown PVC to be capable of propagating fire, once it is ignited, if the cables represent a packing density of more than 1 kg PVC/m.

In addition to the corrosive and choking effect of the HCl gas, opaque black smoke is emitted even before the insulation catches fire. Because the smoke blocks out all light, any escape or exit signs are obscured and the environment is one of total darkness. The production of dense smoke is in many respects more serious than the fire itself. It hinders escape and adds to the difficulty of rescue and of locating and dealing with the fire. To be in opaque and stifling blackness can cause terror which suddenly becomes panic. Logical action and thought are then no longer possible. Those who have been put through a fire fighter's course or a naval damage control exercise will know the feeling.

Cables with improved fire characteristics are now increasingly regarded as essential in enclosed spaces. When PVC burns it generates about 25% by weight of HCl gas. Special fillers can be introduced to reduce the gas evolved but this can increase its flammability: fillers which reduce both HCl emission and flammability, however, lower the mechanical properties.

During tests to assess the durability of signalling cables in mine fires it was found that cables which produce HCl gas, i.e. the halogenated compounds, cease to carry signals before more combustible cables without a halogen content. This is evidently because the acid gas is a conducting medium which is able to short out the cable cores. Again the most suitable type of cable must depend on the application.

4.2 Mineral insulated (MI) cable

A patent taken out by a Swiss engineer in 1896 was for a cable enclosed in a metallic sheath and insulated with inorganic material. The idea remained dormant till 1934 when mineral-insulated copper-sheathed cable (MICC) began to be manufactured in France. Constructions of this type, with either aluminium or copper cores and stainless steel or copper sheaths have a refractory property which makes them a prime choice for fire alarm systems and other circuits, such as emergency lighting, which must continue to function after the outbreak of a fire.

Their method of construction is similar to that of old-fashioned seaside rock with the name of the town running through it. That is to say, it starts as a short, relatively large diameter bar which is drawn down to successively smaller diameters. During the process, the spacing of the components remains in the same ratio to each other so that required clearances between conductors and the sheath are maintained. The mineral insulant is magnesium oxide (MgO) in a highly compressed state in the final cable. These MI cables are totally incombustible and non-ageing. Bare MICC cables can operate continuously with sheath surface temperatures up to 250°C unless they are running alongside or are laid on a material which will not withstand such temperatures. If buried underground or installed with part of the route through a corrosive atmosphere it is sensible to have a PVC oversheath. The surface working temperature then has to be limited to 70°C. For general purposes and where skin burns are not a hazard, a working temperature of 90°C is recommended practice.

Above 250°C the copper begins to oxidise but in principle these cables can continue to function, in a fire for instance, for short periods at around 1000°C.

The main disadvantage with MI cable is that the magnesium oxide is hygroscopic and, unless sealed from the atmosphere at every termination, the insulant slowly turns to the hydroxide $Mg(OH)_2$, which is a conductor. Various types of seal have been designed for this class of cable but most use organic materials and have a limited voltage breakdown strength. Switching spikes which can be produced by the control gear in fluorescent and other discharge lighting fittings can have peak values of several thousand volts, capable of arcing across the small clearances in MI cable seals. The weakness can be overcome by means of glass seals but this requires a more costly technique. If a low resistance is found on an installation with mineral insulated cables, it is advisable to look for a faulty seal. Once this has been repaired the insulation value will usually recover over a period of time. From experience, low insulation readings on newly installed MI wiring do not always indicate the existence of a faulty seal: it can sometimes be the result of the electrician concerned having sweaty hands when the seals were made off.

4.3 Silicone rubber cables and wiring

For high temperature service, connecting into domestic radiant heaters for example, the primary insulation has in the past often consisted of clay beads threaded on to the wires. A preferable solution nowadays is to use silicone rubber. Although considerably more expensive than natural rubber it has quite exceptional properties. SR insulation has a working temperature of 150°C and can withstand brief periods at 200°C. Furthermore, if an SR cable is totally destroyed by fire, it can still function as a cable and if undisturbed will continue to withstand 500 V between cores. Not only has it a high temperature capability, it also can be used down to − 55°C and retains flexibility down to − 70°C. The material is soft and easily cut—primary insulation can be stripped from the wire with the thumb nail. SR wiring is therefore normally protected by a metal enclosure or conduit and is extensively used in electric cookers.

4.4 Cross-linked polyethylene cables

Polyethylene (PE) and polyvinyl chloride both have a 70°C working temperature limitation but PE has better cold flexibility and can be installed at temperatures down to − 60°C. Having no halogen content, it burns without producing acid gases. It also has a lower dielectric loss factor than PVC and so is more suitable for the higher voltages and certainly for higher frequencies.

As its name implies, polyethylene is made by polymerising the gas ethylene $H_2C{:}CH_2$. It has been adopted for power cables mainly in N. America, whereas in European countries PVC has dominated the market.

Since the development of irradiated or cross-linked polyethylene (XLPE) the situation has changed significantly. XLPE can be used at temperatures up to 90°C and is therefore superior to PVC on all counts except perhaps its flammability. This factor is of less importance when the cable is armoured. Raising the operating temperature from 70 to 90°C is particularly valuable in hot environments such as outdoors in tropical countries and in ships' engine rooms as the derating factor for the high ambient will be much less. It seems likely that XLPE will take over more of the market still dominated by PVC and will also continue to replace paper cables over which it has a slight edge with respect to current rating on continuous duty. In addition, XLPE cables permit somewhat tighter bends than equivalent paper cables and, most importantly, do not require a vapour-proof sheath with sealed joints and terminal boxes. Their installation costs are accordingly much less and the simpler techniques reduce the need for specialised skills.

4.5 Thermal ratings

Cables are seldom the cause of a fire unless they are damaged. To start a fire on its own a cable must either suffer an internal electrical fault or be severely overloaded. In either event, if the result is a fire, the fuse or circuit-breaker in charge must also have failed to perform its essential function. The time/current tripping characteristic of the protective device is critical since it must ensure that a short-circuit or earth fault current is stopped before irreversible damage is done to the cables carrying the fault current. It must also prevent overload current from flowing long enough to allow dangerous overheating of the cables to take place. At the same time it must not trip the supply so abruptly that direct-on-line motors with high inertia loads cannot be started.

To enable protection devices to meet these criteria, cables are given two thermal limitations by the manufacturer: a maximum permissible continuous temperature and a maximum permissible short time peak temperature. For example XLPE cable has a working temperature of 90°C and a peak limitation of 250°C. The current which will produce the specified working temperature cannot be stated by the manufacturer since it depends on the thermal conditions of use, namely, the cooling or thermal insulation and the ambient temperature surrounding the cable when it has been installed. The correct current overload setting protecting the cable will be governed by the most adverse of the thermal conditions along the route of the cable.

The peak temperature is normally associated with a fault current rather than overload and, with suitable protection, will not persist for more than a second or so. Any cooling or lagging effect of the cable environment can therefore usually be ignored. The heating effect of a downstream fault will thus be approximately adiabatic with temperature rise determined chiefly by the thermal inertia of the conductors. The rise will then be proportional to the square of the fault current I and its duration t; so the critical parameter is the let-through energy I^2t of the protective device before it trips. If the fault current is not constant, the adiabatic temperature rise in the cable will be given by

$$k \sum I^2 \cdot \delta t$$

during the period of fault current. For practical purposes fault currents can be assumed to be a constant value equal to the prospective short-circuit current PSCC. Lower values will then err on the safe side.

Although the rate of heat loss can be ignored in the case of short-circuit faults, the initial cable temperature will determine the permissible temperature rise to reach the specified peak temperature and may need to be taken into account for the I^2t setting of the protective device. Both the long term and short term thermal limits for cable insulants are empirical values which cannot be found by calculation; nor can the appropriate maximum working temperature be established by any single short term test. What is beyond doubt is that, the higher the working temperature of a cable, the sooner it is likely to fail and a few degrees up or down will profoundly influence its working life.

4.5.1 Method of test

The effect of temperature on the failure rate of the dielectrics used in cable core insulation can be exploited to obtain a suitable thermal rating without having to subject samples of the material to indefinitely long life tests at trial temperatures. The method consists of the application of durability tests at over-temperatures which are hot enough to produce a degradation of the insulation within a reasonable period of time.

The values of the property selected are plotted against time-to-fail for a number of samples, at various temperatures. If the lowest temperature used takes too long to reduce the samples to perceived failure, the plotted values can be extrapolated to failure time provide this further course of the curve is predictable. A typical result is shown in Figure 4.1.

The time-to-fail is determined for at least three different over-temperatures, with sufficient samples in each case to establish reasonable confidence limits. These times are then plotted, together with their confidence limits, against the respective exposure temperatures and extrapolated down to a lower temperature which will give the length of life required, as shown in Figure 4.2. It is recommended that the times are plotted as ordinates on a logarithmic scale and the temperatures as abscissae expressed as reciprocals of the absolute temperature $[T^{-1}]$, where

$$T = 273 + \theta \text{ in degrees Centigrade.}$$

According to the principles concerning the rate of chemical reactions, this form of presentation of the results should produce a straight line graph and so simplify extrapolation to the required failure rate. Because the maximum temperature is unlikely to be maintained continuously when in service, a time-to-fail of three years, or 25 000 hours can be regarded as an acceptable basis for the determination of maximum rated temperature. Periods of use with reduced ambients, with no current, with no voltage, or with part loads should then ensure a reasonable length of life without failure.

Properties whose deterioration can be measured and used as failure criteria in the above thermal tests can be

- Loss of weight
- Contraction or shrinkage
- Mechanical properties such as; flexibility, bending effects, impact strength, tensile strength

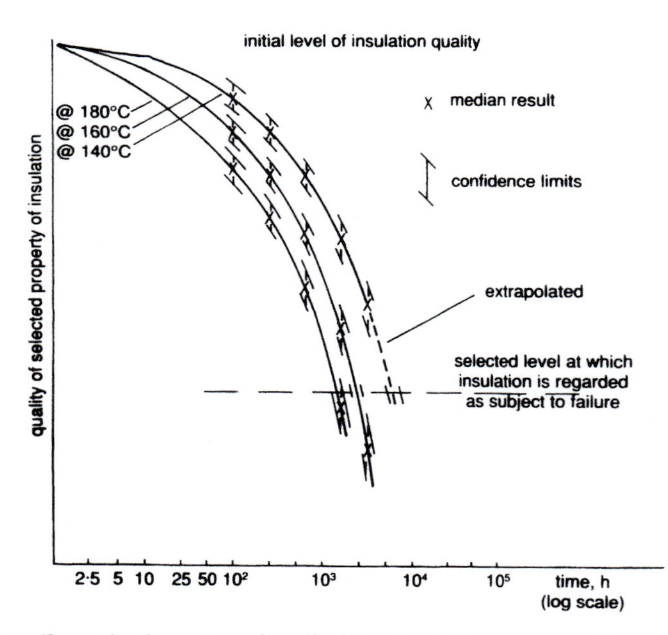

Figure 4.1 *Example of reduction of quality of an insulation property with time at 3 different temperatures*

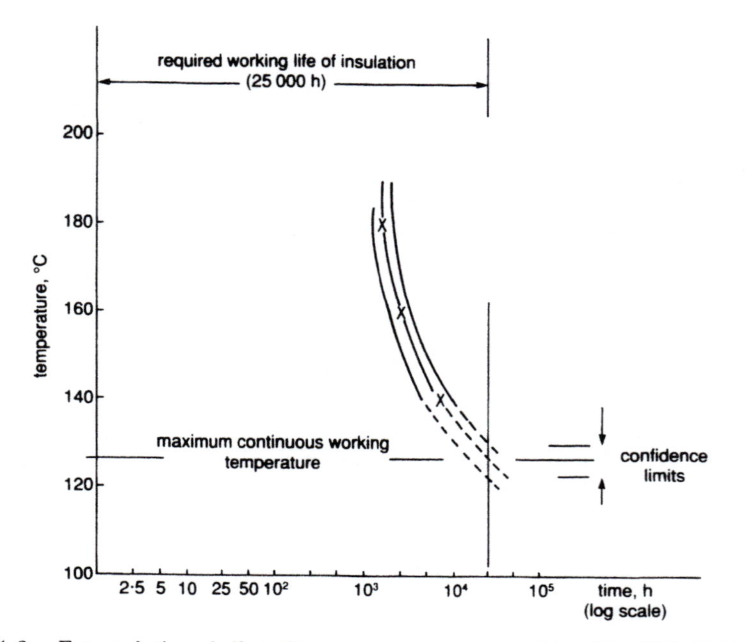

Figure 4.2 *Extrapolation of effect of temperatures to give a working life of 25 000 hours — results based on example in Figure 4.1*

- Voltage withstand strength
- Specific resistivity
- Dielectric loss

It is sensible to select one mechanical and one electrical property for measurement. If a reasonably straight line cannot be obtained by the methods described, it will be sensible to include more than two properties.

4.6 Bibliography

EVANS, B.B.: *Natural rubber, synthetic rubber and thermoplastics for electric cables*; paper presented on behalf of the Cable Makers' Association at the thirty second Convention of the Association of Municipal Electricity Undertakings of South Africa, Cape Town: 1958: Publ. Distribution of Electricity, March 1959.

Papers presented at the first Electrical Research Association Distribution conference, Edinburgh: 1967.

BONIKOWSKI, Z.: *Cables for hazardous environments*; Electronics and Power, Vol.29, No.4, pp 317–320, Publ. IEE., 1983.

SARNEY, R.G.: *Cables for the offshore oil and gas industry*; Electronics and Power, Vol.31, No.4, pp 304–306, Publ. IEE., 1985.

FINLAYSON, A.J.: *Low smoke and fume cables for BNFL Sellafield Works*; Power Engineering Journal, Vol.4, No.6, pp 301–307, Publ. IEE., 1990.

Recommendations for testing solid insulating materials for assessment of their thermal stability; VDE 0304 [Engl. ed'n 1965], Publ. VDE-Verlag GmbH, 1 Berlin 12.

Chapter 5

Electrical equipment for use in explosive atmospheres

With the wider use of natural gas and of petroleum accidental ignitions of accumulations of flammable gases and vapours are a major industrial hazard. The mode of combustion is usually explosive and often results in injury, death and widespread destruction. In today's complex society the only wholly safe procedure when gases and vapours are released to the atmosphere is to ensure that they are immediately ignited. This may sound anomalous but the safest way to vent a flammable gas is to burn it off at the point of release. A flare stack works on this principle. A processing furnace using an oxygen-free atmosphere and gas such as hydrogen for instance can be protected from dangerous release of gas by a so-called flame curtain consisting of small flame jets round the furnace portals.

The most hazardous conditions arise when gas escapes unburnt and mixes with air to form a very large volume of explosive mixture. On ignition, the energy released will create a sudden increase in temperature before the mixture has time either to expand or to lose heat to the surroundings. This increase in temperature from say 280 K to over 2000 K will, if it takes place in the open, produce an expanding fireball followed by a partial vacuum as the hot gases rapidly cool down. The suction effect may not destroy buildings but can pull out nearby windows. In a confined space an adiabatic temperature rise of eight times the absolute temperature will produce an eight-fold elevation of pressure. This represents a momentary internal pressure on the walls of the enclosure of seven atmospheres or about 70 tonne/square metre. Obviously no normal building will withstand this and the usual effect is that the ceiling lifts, the walls go outwards and the ceiling then collapses on the occupants.

Explosions of combustible dusts are also highly destructive. Most coal-mine disasters have been caused by a combination of gas and dust. An initial methane explosion at the coal face produces a pressure wave along the coal face and out-by down the mine roadways, raising and igniting clouds of deposited coal dust as it travels. The flame front thus gathers fuel and heat as it travels through the mine workings.

Conversely, in the case of oil well ignitions, the fuel is fed into the seat of the fire by the geostatic pressure of the well or, as in the Piper Alpha North Sea catastrophe, by the pressure in the long undersea pipeline.

In all these cases the prevention of uncontrolled or undetected escape of flammable gas or vapour is paramount. Where this cannot be ensured the area is termed a Hazardous Area involving total or conditional exclusion of any possible sources of ignition. This presents a considerable restriction on work and installations permissible in the area concerned. Accordingly increasing attention has been given over the last fifty years to a process known as hazardous area classification, or HAC, by which three different levels of hazard can be allocated.

5.1 Development of area classification

Hazardous area classification is based upon the probable frequency and the likely maximum duration of a flammable atmosphere being present.

The probability of an explosive mixture occurring at any point on an installation cannot be precisely determined and, in any case, will vary with time. For instance, it will increase with the age of the plant and whenever the equipment is being started up or shut down. It may also increase during instrument failure, power failure or exceptional weather conditions. Most importantly, it will be affected by the plant maintenance and replacement policy.

The lowest probability of accidental discharge of flammable gas or vapour will usually be under steady state, designed load conditions when no maintenance is in hand; yet adequate maintenance is the most important single factor in reducing risk of fire and explosion in the long term.

The boundary of a hazardous zone is where the dilution of the flammable gas with air can confidently be expected to bring the explosive mixture below the lower explosive limit (LEL). For this part of the classification exercise the values of several important parameters have to be established or assumed. These will include the maximum rate of release of gas and the minimum wind velocity since the lower the wind speed the larger the hazardous area to be expected for a given rate of release.

The International Electrotechnical Commission (IEC) has defined the three levels of hazardous area as follows:

Zone 0 An area in which an explosive gas mixture is present continuously or is expected to be present for long periods, or for short periods which occur frequently.

Zone 1 An area in which an explosive atmosphere can be expected to occur periodically or occasionally during normal operation.

Zone 2 An area in which an explosive atmosphere is not expected to occur in normal operation and if it occurs is likely to be present only infrequently and for short periods.

NOTES: It is advisable to consider concentrations above the upper explosive limit (UEL) as an explosive atmosphere for HAC purposes.
Catastrophic failures such as the bursting of pressure vessels or pipelines are not considered to be sources of release for HAC purposes.

These two notes taken together mean that the inside of a pipeline or vessel should generally be considered as Zone 0 and that the outside of a welded pipeline or closed vessel can be regarded as a safe area—so that it is possible to accept a single vapour barrier (the wall of the pipe or vessel) between Zone 0 and safe areas.

The British Standard, BS 5345 Part 2: 1990 'Classification of hazardous areas' has adopted the IEC definitions. American practice, which is based on Article 500 of the National Electrical Code of the National Fire Protection Association (of the USA), differs from European practice in that the areas are classified into Divisions 1 and 2, Division 1 embracing both Zones 0 and 1, as defined above.

As stated, HAC depends upon the probable frequency and duration of an explosive gas/air mixture being present. No definite figures have been universally agreed but some fairly widely accepted values are shown in Table 5.1.

Table 5.1 Recognised probability limits for hazardous areas

IEC classification	Safe area	Zone 2	Zone 1	Zone 0
NEC classification	Safe area	Division 2	Division 1	Division 1
Probability of an explosive atmosphere	<0·01%	0·01 – 0·1%	0·1 – 10%	10 – 100%
Hours/year	<1	1 – 10	10 – 1000	>1000

The resulting HAC affects the type of electrical equipment that may be used but this is not necessarily a complete solution. Even in controlled hazardous areas there are generally other more likely causes of ignition than electrical apparatus. It may be better to modify the process side of the installation and reduce the hazardous areas rather than place undue restrictions on the design, operation and maintenance of the electric power and instrumentation systems. It is being increasingly recognised that HAC is not the proper function of the electrical engineer alone but that it should be undertaken by a team. This should include the factory or process engineer, the operating and maintenance engineers and the site safety officer.

The first stage is to make an inventory of all possible points of escape capable of creating an explosive atmosphere of dangerous proportions. These possible sources of release are then graded into three levels according to the probability of an escape taking place, namely

Continuous grade sources of release, i.e. sources which will release flammable gas or vapour continuously or for long periods, or frequently for short periods.

Examples – The surface of any flammable liquid at a temperature above its flash point, unless it is in an inerted vessel (see Section 13.3.2)
– The liquid surface of an oil/water separator
– Vents which release gas or vapour to atmosphere frequently or for long periods
– Extraction ducts for flammable solvent vapours.

Primary grade sources of release, i.e. sources which can be expected to release flammable gas or vapour periodically during normal operation.

Examples – Many types of reciprocating and rotating seals
– Water drains for vessels containing flammable liquids
– Sampling points.

Secondary grade sources of release, i.e. sources which are not expected to release in normal operation and are likely to do so only infrequently and for short periods.

Examples — Rotating and reciprocating seals which are not expected to release gas or vapour during normal operation
— Flange couplings for pipes etc. and other bolted seals
— Pressure relief valves and other openings which are normally closed
— Sampling points which are not expected to release flammable material to atmosphere in normal operation.

Continuous grade sources of release will normally create a Zone 0. Primary grade sources of release will normally create a Zone 1. Secondary grade sources of release will normally create a Zone 2. However, there are exceptions to these rules. Where the released vapour or gas is unlikely to be dispersed because there is insufficient ventilation, or it becomes trapped — in a ship's hold or in a cable trench for example — a more hazardous zone will be created. This is because under stagnant conditions there will be a higher probability of a residual explosive atmosphere. Figure 5.1 shows the example of an unventilated cable trench in the vicinity of a Zone 2 area.

5.2 Assessment of extent of zones

Having graded all possible sources of release it is then necessary to estimate the extent of the resultant hazardous areas. These estimates will need to take several factors into account, in particular

- Density of the flammable gases (heavier or lighter than air)
- The lower explosive limit
- The amount of ventilation
- The pressure at the source of release
- The equivalent aperture at the point of release
- The slope of the ground and drainage.

Points to note:

(a) If the gas released is heavier than air it will spread along the ground. The hazardous zone will thus have a greater area at ground level. Most gases which

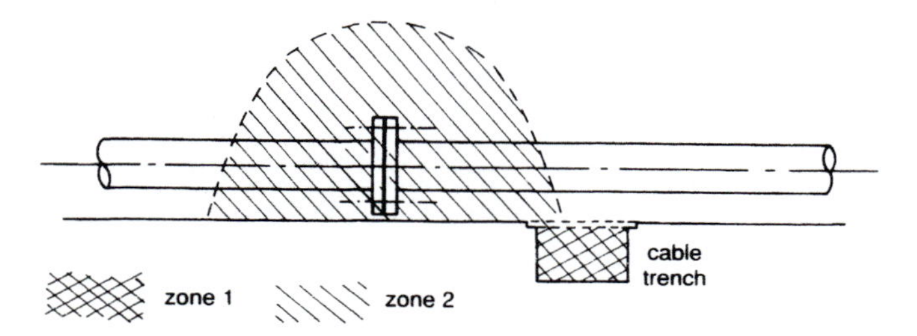

Figure 5.1 Example of a secondary grade source of release (flange coupling) producing both Zone 2 and Zone 1 hazardous areas

are explosive and all flammable vapours are heavier than air. Accordingly they can lie indefinitely in drains and trenches and, in windless conditions, on the water surfaces of enclosed docks. These vapours can also flow considerable distances beyond the assumed hazardous area and when ignited in the presumed safe area will flash back to the source of release. Figure 5.2 shows typical HAC for continuous, primary grade and secondary grade sources of release for a flammable vapour.

(b) The lower explosive limit (LEL) is the 'leanest' mixture with air which can be ignited at normal temperature and pressure. Figure 5.3 shows the limits of ignitability of various gases in air by volume. Gases with lower values of LEL will tend to form larger hazardous areas when released.

(c) Ventilation assists the dilution with air down to a level below the LEL and so will reduce or even prevent the creation of a hazardous area. Where operators are working within a building high air velocities can cause discomfort. Up to 0·1 m/s at 16°C and up to 0·3 m/s at 24°C are acceptable in this context. Within an unventilated enclosed room hazardous and safe areas cannot both exist. It must be either totally safe or totally hazardous.

(d) The release of gas at a high pressure will cause turbulence and rapid mixing with air to dilute the gas to below the LEL. If the escaping jet is impeded by an obstruction, such as pipe lagging, so that it disperses into a fairly still air the hazardous zone can be much greater. The release of liquid at high pressure will produce a jet of liquid breaking down into a mist and possibly also forming a pool, so creating two centres of hazard some distance from the source. Figure 5.4 indicates the effect of a leak in a high pressure gasket.

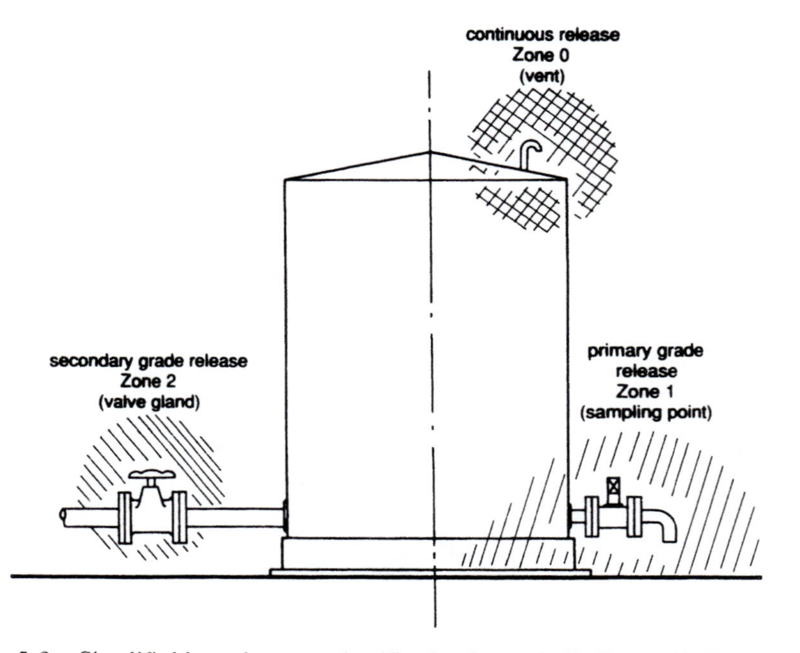

Figure 5.2 Simplified hazardous area classification for a volatile flammable liquid such as petrol (based on the codes of practice BS 5345 Part 2 and IEC Publ. 79-10)

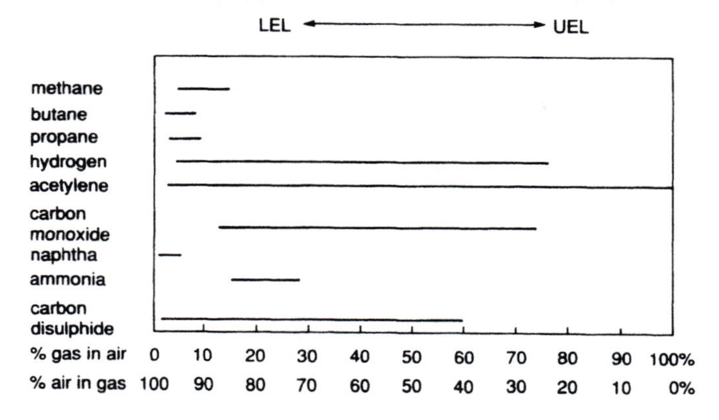

Figure 5.3 Range of flammability of various gases in air, by volume

The energy required to ignite each gas is a minimum about midway between its upper explosive limit (UEL) and its lower explosive limit (LEL)

(e) The rate of release of a product under pressure will be a function of the equivalent size of the aperture. Except in the event of a catastrophic failure, the most important points of escape under pressure will generally be the result of failed sealing materials and devices, the rupture of instrument bellows, capillary lines etc. At normal temperature and pressure the vapour phase has about 1000 times the volume of the liquid phase. This means that if the LEL is 2% for example, a volatile liquid on release can produce up to 50 000 times its volume of explosive atmosphere.

(f) As mentioned, heavier than air vapours will tend to flow into ducts and trenches. The slope of the ground must be considered especially if flammable liquids are released or spilled.

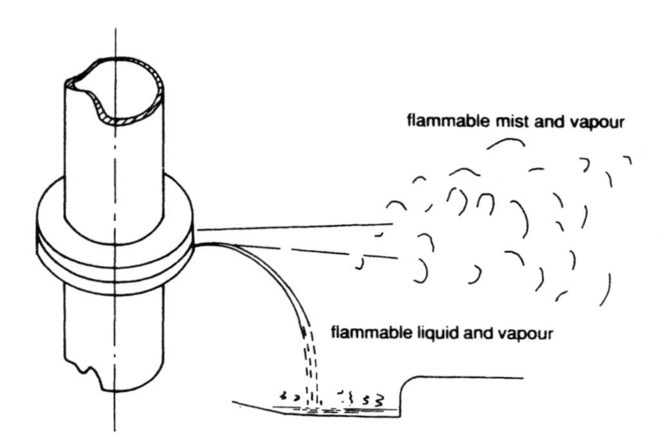

Figure 5.4 Release of high pressure flammable liquid in the form of a jet

5.3 Other properties of explosive atmospheres

Besides the above considerations, there are three other important properties of flammable gases which significantly influence the potential danger although they do not affect the size of the hazardous area. These are the flash point, the ignition temperature and the required ignition energy.

5.3.1 The flash point

This is the temperature of a liquid at which its vapour pressure is high enough to produce a flammable atmosphere (the LEL) immediately above its surface. Flash points above ambient temperature can be determined by warming the liquid in an Erlenmeyer flask and applying a lighted taper, as indicated in Figure 5.5*a*. While the liquid is below its flash point, no flammable atmosphere will be formed. Liquids below their flash point cannot produce a hazardous area unless they are ejected as a mist or are allowed to impinge on a hot surface.

5.3.2 The ignition temperature

This is the lowest temperature at which a gas or vapour will spontaneously ignite with air. As indicated in Figure 5.5*b*, at the ignition temperature the gas/air mixture explodes without the application of any additional source of ignition such as a lit taper. It is obviously a higher temperature than the flash point and governs the maximum safe temperature for any heated body or surface which can be

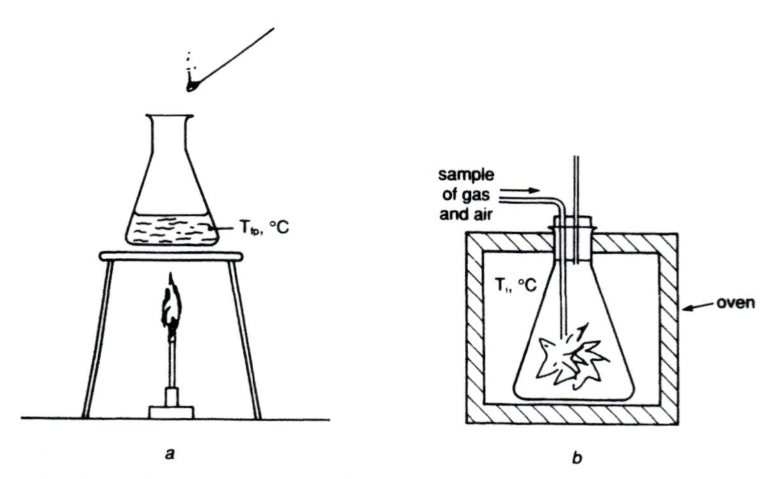

Figure 5.5 The difference between flash point and ignition temperature

At the flash point of the liquid, sufficient vapour is produced to form an ignitable mixture with air. At the ignition temperature, the sample of gas and air will explode spontaneously after being introduced into the Erlenmeyer flask. T_i is regarded as an unsafe free surface temperature in the presence of the gas

a determination of the flash point T_{fp}
b determination of the ignition temperature T_i

allowed in a hazardous zone. It may be noted here that, with electrical apparatus, peak surface temperatures are often remarkably difficult to determine with accuracy, e.g. the rotor bars of a cage motor under starting conditions.

5.3.3 The minimum ignition energy

This is the lowest electrical spark energy required to cause ignition under standardised conditions. It bears no relation to the ignition temperature, some gases with a high ignition energy having a low ignition temperature and vice versa. The repeatability of this test is poor because of the variable nature of the energy transferred from the spark to the gas, but it is an important property for the method of protection of hazardous area electrical apparatus known as intrinsic safety (discussed later in Chapter 7).

Values of the above-mentioned five properties for some typical gases and vapours are shown in Table 5.2. It will be noted that the safe spark energies for these are well below 1 mJ, i.e. less than 1 W for 1 ms. In mechanical terms this is less than 1 g falling 10 cm.

5.4 Electrical sources of ignition

Figure 5.6 shows the approximate temperatures generated by various types of equipment compared with the range of temperatures required to ignite flammable gaseous mixtures. It will be seen that the working temperatures of most electrical insulation and consequently the surface temperatures of most insulated conductors are below the ignition temperatures of nearly all explosive gas mixtures. Arcs and sparks on the other hand are highly incendive. However, besides the obvious items such as sparks from switch contacts and from welders and the higher temperatures of electrical heaters and quartz lamp bulbs, electrical apparatus can have a number of less self-evident incendive properties. These include

- incandescent particles emitted through the gaps in flameproof equipment
- sparks produced in nominally non-sparking equipment due to internal circulating current—particularly in high voltage cage motors
- static discharges from plastic covers and enclosures
- sparking from inductive and capacitive pick-up on unused cable cores
- chemical ignition from discarded sodium vapour lamps
- HV and RF discharges.

Table 5.2 Values accepted in the UK for some typical gases

	Hydrogen	Pentane	Methane	Kerosene
LEL% V/V	4·0	1·4	5	–
Flash point °C	gas	< –20	gas	+38
Relative density (compared to air)	0·07	2·48	0·55	–
Ignition temp. °C	560	285	595	210
Mininum ignition energy (permissible spark energy in mJ)	0·02	0·2	0·2	–

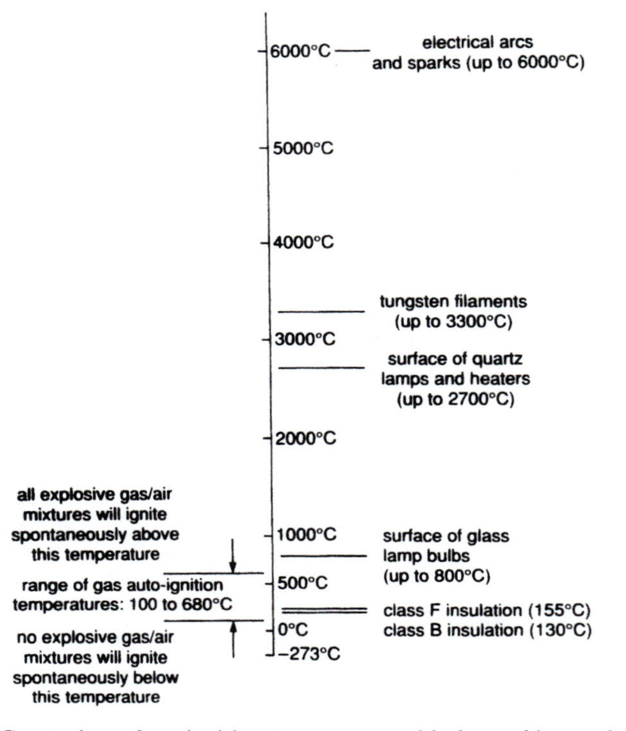

Figure 5.6 *Comparison of gas ignition temperatures with the working surface temperatures of electrical apparatus. Class B insulation in normal use is capable of igniting only carbon disulphide and Class F insulation only CS_2 and acetaldehyde*

Many of these dangers can be greatly reduced by suitable design, good installation practice and skilled operation and management. The apparatus design is subject to the requirements of the standard for the type of apparatus concerned and also the standard for the ignition protection concept which is adopted. In the UK, installations in explosion-hazardous areas and their subsequent operation and maintenance are subject to the IEE Wiring Regulations and also to the BS Codes of Practice concerning electrical installations in potentially explosive atmospheres.

Electricity is not of course the only or even the main source of ignitions. Compliance with the product standard, the flammable atmospheres protection concept standard, the Wiring Regulations and the flammable atmospheres installations Code of Practice does not therefore deal with all possible causes of ignition which can be attributed to the use of electrical equipment. For example, an electrically powered forklift truck could set off an explosive atmosphere in any of the following ways:

Electrically

in normal operation	by faulty operation
— sparking from motor brushes	— sparks from battery
— sparks from contactors	— exposure of light filament
— hot resistor surfaces	— faulty wiring
— sparks from switches	— faulty solenoids

Mechanically

in normal operation	by faulty operation
— heating of brake shoes	— overheated bearings
— frictional sparking (e.g. steel forks on concrete)	— falling loads

Assessing the suitability of electrical apparatus for use in hazardous areas is usually undertaken by an independent test house which will then issue a certificate showing compliance with the protection concept concerned. This will cover the relevant electrical aspects of the design but, as can be seen from the above example, the certificate will not necessarily cover all the risks which may arise in practice.

5.5 The design of electrical apparatus for use in hazardous areas

There are at present eight standardised protection principles for the safe use of electrical apparatus in potentially explosive atmospheres, in addition to some non-standardised methods. The standardised principles are commonly referred to as types of protection and, for each of them, harmonised European standards, or Euro-norms (ENs) have been issued. The first batch appeared in 1977 since when they have been under constant review in the light of experience. Each type of protection, the apparatus marking codes and the relevant standards with which the apparatus must comply are shown in Table 5.3. The letter E may precede the Ex mark to indicate compliance with a Euro-norm.

Each of the standards and codes in Table 5.3 applies to a broad range of electrical products. They are accordingly termed horizontal standards. Equipment which by its nature has to operate at high voltages or temperatures in explosive atmospheres has specific standards and codes and British Standards of this nature include:

BS 6742 Parts 1, 2, 3 and 4 for electrostatic spray guns
BS 6351 Parts 1, 2 and 3 for electric surface heaters, trace-heating cables in particular
BS 4533 Part 102–51 for light fittings (luminaires) intended for use in Zone 2 areas.

In addition, BS 7117 Parts 1, 2 and 3: 1991 deal exclusively with petrol pumps for garage forecourts.

5.6 General principles of design

With very few exceptions, the ignition of a flammable atmosphere results in a potentially destructive explosion, caused by the heat generated in a thermochemical

Table 5.3 Types of protection and codes of practice (CoP)

Oil immersion [UK CoP: BS 5345 Part 9]
Marking code: Ex o
Certification standard: BS 5501 Part 2/EN 50015
Corresponding IEC Publ.: IEC 79-6
Sparking and/or non-sparking apparatus is immersed in oil so that it cannot ignite gas above the oil surface.

Pressurised apparatus [UK CoP: BS 5345 Part 5]
Marking code: Ex p
Certification standard: BS 5501 Part 3/EN 50016
Corresponding IEC Publ.: IEC 79-2
The apparatus is in an enclosure containing air or other inert gas at a positive pressure to prevent the ingress of an external explosive atmosphere. The method also includes the continuous introduction of inert gas into an electrical enclosure containing a source of emission of explosive gas.

Powder/sand filling [UK CoP: none *pro tem*]
Marking code: Ex q
Certification standard: BS 5501 Part 4/EN 50017
Corresponding IEC Publ.: IEC 79-5
The electrical enclosure is filled with heat resistant powder or granules so that, if an arc occurs, it will not ignite an external explosive atmosphere. Used for apparatus up to 6600 V with no moving parts.

Flameproof enclosure [UK CoP: BS 5345 Part 3]
Marking code: Ex d
Certification standard: BS 5501 Part 5/EN 50018
Corresponding IEC Publ.: IEC 79-1
The electrical enclosure is designed to withstand an internal explosion of specified gas mixtures without damage and without igniting the external gas mixture. This method does not protect the electrical apparatus itself from damage due to an internal explosion.

Increased safety [UK CoP: BS 5345 Part 6]
Marking code: Ex e
Certification standard: BS 5501 Part 6/EN 50019
Corresponding IEC Publ.: IEC 79-7
Non-sparking apparatus having increased reliability, e.g. by derating and by non-loosening connections, to prevent incendive faults occurring during the operation of the apparatus.

Intrinsic safety [UK CoP: BS 5345 Part 4]
Marking code: Ex ia and Ex ib (ia is safer than ib)
Certification standard: BS 5501 Part 7/EN 50020
Corresponding IEC Publ.: IEC 79-11
The restriction of maximum possible energy supplied to the apparatus, to below an incendive level. For freely interconnected apparatus, the restriction of energy must be applied to the whole system.

Encapsulation [UK CoP: none *pro tem.*]
Marking code: Ex m
Certification standard: BS 5501 Part 8/EN 50028
Corresponding IEC Publ.: —
Parts which could ignite an explosive atmosphere are sealed within a compound such as moulding plastic.

Intrinsically safe systems [UK CoP: BS 5345 Part 4]
Marking code: Ex ia or Ex ib
Certification standard: BS 5501 Part 9/EN 50039
Corresponding IEC Publ.: IEC 79-11
Requirements for interconnected intrinsically safe installations—in particular, where only part of the system is within a hazardous area.

Type of protection N [UK CoP: BS 5345 Part 7]
Marking code: Ex N
Certification standard: BS 6941/(EN 50021 in draft)
Corresponding IEC Publ.: IEC 79-15
Requirements for the protection of apparatus for use in Zone 2 areas, i.e. in less hazardous areas. The marking code will be Ex n in the harmonised Euronorm.

Special protection [UK CoP: BS 5345 Part 8]
Marking code: Ex s
Certification standard: to test house requirements
Corresponding IEC Publ.: —
Non-standardised methods of protection shown by test to be sufficient for the equipment concerned to be safely used in hazardous areas.

The rules for electrical apparatus used in flammable dust atmospheres vary from one country to another. In the UK, BS 6467 Parts 1 and 2 apply for the protection of such apparatus by enclosure and BS 7535 applies for other types of protection against dust explosions.

chain reaction. In this respect, the LEL of a gas/air mixture is analogous to the critical mass of a nuclear device. Hence electrical apparatus for use in potentially explosive atmospheres must be designed so that it is not capable of initiating an explosion. This would be relatively simple if it were merely a matter of excluding or enclosing normally sparking and heated components. Abnormal operation, however, must also be considered since electrical faults and mechanical overloads can often lead to an incendive condition. The design requirement is therefore, either to prevent these faults from occurring, or to ensure in some way that they cannot cause ignition.

The first designs for this type of equipment were for use in coal mines where early enthusiasm for electric power gave rise to many fires and deaths. At first the dangers were neither realised nor properly understood and a six-volume treatise on 'Modern electric practice', published in 1909, stated

> 'A great deal has been said about the danger of using electricity as a motive power in fiery mines, but this danger has been greatly exaggerated. It must be remembered in connection with this point that, before sparking, even if it occurs, can become dangerous, an explosive mixture must exist at the place where the spark is formed. Such an atmosphere would never be found in any well-regulated mine, unless caused by a sudden outburst of gas.'

Within a decade or so, the many fires and deaths in coal mines had reversed this cavalier attitude. The need to prevent tampering with the flameproof enclosure resulted in British flameproof equipment being fastened with bolt heads and nuts in counterbored recesses whereas, previously, terminal boxes in mines had been secured with wing nuts.

Until 1912 it had been assumed that circuits fed from primary (Leclanché) cells would not be capable of igniting firedamp. In that year the worst mining disaster ever to occur in the UK took place at Senghenydd Colliery, near Cardiff. This led to an investigation into the minimum prospective fault currents capable of igniting methane. It was found that the energy in the inductance of the very low voltage signalling bells used in the mines was a potential hazard but that, below certain voltages, currents and inductances, specified atmospheres would not be ignited.

Circuits where these safe fault levels cannot be exceeded are termed intrinsically safe circuits and this method of protection has over the last 60 years been greatly developed, largely in the UK. It has become particularly important with the expanding application of low power, low voltage solid-state systems for process control and information technology. Flameproof enclosures for power equipment and intrinsic safety for light-current apparatus are still the two most important and widely used concepts for electrical installations in explosion-hazardous areas and will be considered in further detail in Chapter 6 and Chapter 7 respectively.

5.7 Harmonised standards and the New Approach

In the past the need for certification of hazardous area apparatus created a technical barrier to trade because each country used its own national certification standards. The committees of the Common Market electrical standards organisation (CENELEC) have accordingly been required to produce harmonised standards (ENs) to prevent goods being certified by one member state from being rejected by another on professed grounds of insufficient safety. The harmonisation process with respect to hazardous area apparatus has still to be completed and must now be regarded as very time-consuming and somewhat intractable. A major disadvantage lies in the continuously increasing complexity of the standards: they are for ever being revised and augmented to reduce ambiguities and differences in interpretation so that each EN has to be read with a lengthening appendage of additional clauses and amendments.

When drafting international standards, a great deal of committee work and time are involved in seeking to protect national commercial interests. BSI committees responsible for hazardous area apparatus still handle over 3000 pages of discussion documents a year and this has continued unabated since BSI joined CENELEC in 1972.

Although these Euronorms are long and detailed, they do not incorporate state-of-the-art technology but merely the highest common factor of agreement in the drafting committees, which necessarily include in their membership the widest possible range of, often conflicting, commercial interests. Furthermore, some duplication of effort has arisen due to the need to respond to separate initiatives by CENELEC and the IEC. A recent decision has therefore been taken in Geneva to co-ordinate the respective IEC and CENELEC programmes of work. A further major harmonisation task is shortly to be imposed on CENELEC and the standards committees of member states. This is in consequence of the proposed EC Council Directive entitled 'The New Approach Directive'. It will require products to comply with a rather different set of criteria from those of the present hazardous atmosphere ENs. The new criteria are to be referred to as Essential Safety Requirements (ESRs). These will first need to be established for each type of protection and then incorporated in the respective harmonised standards.

Under the New Approach, products will be allowed to meet their ESRs in any number of ways, but the preferred way will be by compliance with the appropriate ENs once they have been suitably revised and accepted by the EC Commission. National standards may possibly be approved by the Commission where there are no ENs and it is shown that the standards meet the Essential Safety Requirements.

The proposed Directive is complex and detailed and is intended to cover both Group I (mining) equipment and Group II (non-mining) equipment as well as flammable atmospheres due to either gases or dusts. In essence it specifies three levels of protection, referred to as

Conformity category 1 (for Zone 0 hazardous areas)
Conformity category 2 (for Zone 1 hazardous areas)
Conformity category 3 (for Zone 2 hazardous areas)

Category 1 requires two independent types of protection, or one type of protection designed so that the apparatus remains safe (non-incendive) after two independent faults. Category 2 requires one type of protection designed to ensure the apparatus remains safe with one foreseen possible fault. Category 3 apparatus must normally be non-incendive and a fault producing an incendive condition must be unlikely to occur.

Category 1 is for electrical equipment where a hazardous atmosphere is normally present, e.g. for float switches within a vessel of flammable liquid or gas flow meters within a pipe. It is therefore reasonable for Category 1 devices to be 'doubly safe'. Category 2 is for installations which must remain safe when there is an occasional escape of gas, without it being necessary immediately to de-energise the equipment. Category 3 equipment may be incendive under single fault conditions. These faults must be unlikely, but even so, such equipment will need to be de-energised if it becomes surrounded by a persistent explosive atmosphere.

Types of protection for Categories 1 and 3 should not present new problems. Category 2, however, will need consideration with respect to flameproof enclosures, oil immersion and increased safety since the equipment will have to remain in a protected state with one fault.

It can be argued that the present ENs for these types of protection are insufficient for Category 2 apparatus. Thus, a high power short-circuit within a flameproof enclosure could rupture the case and be a source of external ignition. Similarly, a high temperature lighting fitting will be in an incendive condition with one fault if the glass is cracked or broken. As regards oil immersion, a leak in the tank could drain the oil and expose sparking or over-heated components to the surrounding atmosphere, making it unsafe with one fault. Increased safety is also likely to be unsafe with a single electrical fault. Although the method applies only to non-sparking equipment, a break in the circuit—for example, a broken bar in the rotor of a cage induction motor—is a single fault which would leave the machine in an unprotected sparking or arcing condition.

Chapter 6

Protection by flameproof enclosure

Some electrical apparatus can be hermetically sealed but this is impractical for most industrial electric power equipment. If an enclosure is able to 'breath', an external explosive atmosphere will be able to enter the housing as a result of changes in the internal temperature. Whenever the apparatus is switched off it will tend to suck in some of the outside atmosphere as it cools. This applies particularly to lighting fittings where there can be a large variation between 'on' and 'off' temperatures. It also applies to a lesser extent to electric motors where a running clearance is usually necessary for the drive shaft, even with totally enclosed fan cooled (TEFC) motors.

The principle of flameproof protection is to place electrical equipment in an enclosure which does not need to be sealed but which will not ignite a surrounding explosive gas if the same explosive mixture is ignited within the enclosure. A flameproof enclosure is therefore in effect a type of pressure vessel in which all openings and running clearances have been shown by test to be reliable flame traps.

The early development of flameproof protection took place largely in Germany and in the UK. In 1921 the Mining Department of the University of Sheffield undertook to conduct research work on flameproof enclosures, in collaboration with the Safety in Mines Research Board (now part of the Health and Safety Executive) and in consultation with the British Electrical and Allied Industries Research Association (now Electrical Research Association (ERA) [Technology] Ltd.). The first British Standard on the subject was BS 229, issued in 1926 and Sheffield University then began to issue certificates to manufacturers stating compliance with this standard.

About this time, national testing and certification was established in the USA and on the European mainland. In 1931 the duty of certification for British mining equipment was taken over by the Mines Department of the Board of Trade. In that year a self-contained testing station was set up for the purpose at Harpur Hill near Buxton. This gave rise to the world-renowned term Buxton Certificate, the site still being used by the Health and Safety Executive for the same purpose. Experimental work with explosive atmospheres can be very noisy and this testing station high up in the Pennines is well placed, away from a dense urban population. It may be noted here that, while there have been many refinements, the basic tests have remained the same, namely, a peak pressure determination, an over-pressure test and a flame transmission test.

Initially the Buxton Certificates, also known as FLP Certificates were concerned only with requirements for the mining industry and were made solely with methane/air mixtures. It soon became necessary to establish tests in other flammable gases since flameproof products tested only with methane/air mixtures were in fact being used in industrial premises where other gases or vapours were likely to produce an explosion hazard. The first Sheffield certificate to specify the explosive mixture used for the test was accordingly issued in 1928.

Largely because the original BS 229 standard was based on mining practice and related to heavy duty and mainly cast iron enclosures, it imposed the following rules for flameproof housings:

- All face joints such as flanges to be machined and to have a land of at least 1 inch from the outside to the inside of the enclosure
- Flange faces may form a tight metal to metal joint or they may be held apart to form a vented joint with a maximum permissible gap of 0·02 inches
- Jointing with rubber, asbestos or any such material liable to deterioration was not permissible
- Lengths of bearings for shafts and spindles to be not less than 1 inch, with a maximum radial clearance of not more than 0·01 inch
- Direct entry of cables into the main flameproof enclosure was not permissible
- Nuts and bolt heads securing covers etc. to be shrouded.

Although they have subsequently been modified and considerably augmented, these rules remain the basis of flameproof designs. Figure 6.1 shows an example of a flameproof enclosure to modern certification standards. If a sealing gasket

Figure 6.1 Flameproof enclosure with weatherproofing O-ring

The certified flame path in this case is the cylindrical annulus formed by the spigot

such as an O-ring is used, the joint must be shown to be flameproof both with and without the gasket.

When BS 229 was first published, much of the high voltage switchgear used in mines was of a flameproof oil-immersed type. However, subsequent research showed that the gases produced by the arcing under the oil were mainly hydrogen and acetylene, both of which are far more 'lively' than firedamp/air mixtures. A different and more rigorous set of rules is therefore necessary for oil-immersed equipment protected by flameproof enclosure. As a result, and also because of the flammability of the oil itself, oil-immersed apparatus is no longer considered suitable for flameproof protection. On the other hand, in mainland Europe, oil-immersion alone is regarded as a safeguard so that, although certain tests are carried out for its use in explosive atmospheres, the flameproof tests, as such, are not applied. Accordingly, as previously indicated in Table 5.3, there are separate ENs for Ex d and for Ex o types of protection.

6.1 Principles of testing

The testing of flameproof apparatus requires considerable expertise and investment in costly equipment. It is therefore impractical for any but the largest manufacturers and those specialising in flameproof products to undertake their own prototype testing before submitting a new design for certification.

Among the tests required by BS 5501 Part 5/EN 50018 are

Test (a): to determine the maximum possible pressure which can be caused by a gas explosion within the enclosure
Test (b): in which the enclosure is subjected to a somewhat greater pressure than that determined from Test (a)
Test (c): to show that, again with a reasonable factor of safety, a gas explosion within the enclosure will not ignite an explosive atmosphere surrounding the enclosure.

These tests require some sophistication in proportioning and assaying the gas mixtures and in the measurement of the resulting peak explosion pressures. In addition, the nature of the gases concerned and the fact that the apparatus under test is by definition unproven means that very comprehensive safety measures have to be applied in the management of the test laboratory.

With respect to Test (a), the rate of rise of pressure within an enclosure can be extremely fast. Accurate values for the peak pressure are therefore difficult to obtain due to inertia and mechanical resonance effects in the pressure sensors. The readings can also be affected by the heat of the flame.

Although the explosion may be very rapid and may sound instantaneous, there is in fact a flame front moving through the enclosure which can precompress the gases in front of it, as indicated in Figure 6.2. A 50% precompression, for instance, could cause 50% increase in the explosion pressure. Hence a number of comparative pressure sensing devices may be required in an initial test to detect the position of maximum pressure. Internal pressure wave reflections and the ringing or mechanical resonance of the enclosure itself are further factors which can reduce the factor of safety. The test pressure is normally 1·5 times the peak pressure determined in Test (a).

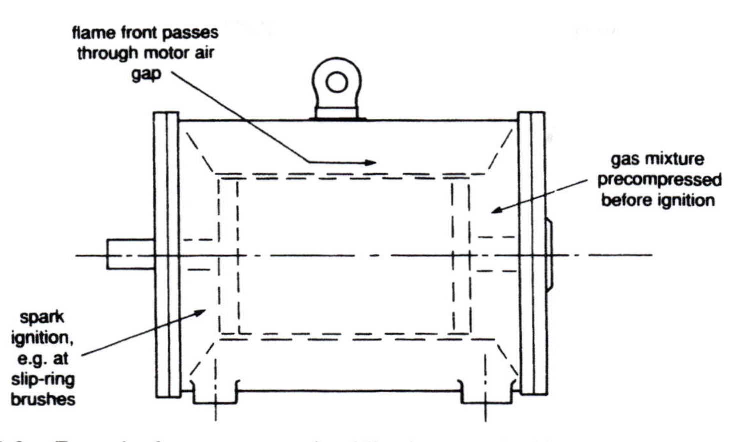

Figure 6.2 Example of gas precompression following a gas ignition at one end of a flameproof motor

With respect to Test (b), the over-pressure can be applied statically e.g. hydraulically with a hand pump, or dynamically, e.g. by repeating Test (a) but precompressing the gas mixture to 1·5 atmospheres before ignition, or by using an oxygen enriched mixture.

With respect to Test (c), which is the critical flame transmission test, a special sample of the flameproof enclosure is required. This must be prepared so that it has the maximum gaps and clearances permitted by the tolerances on the manufacturer's drawings.

One method of carrying out this test consists of placing a plastic film round the test piece. Using a polythene sleeve, it can be roughly sealed by a suitable loop of elastic cord placed round a groove in a circular base-board. The upper end of the sleeve is similarly closed by a small elastic band. The equipment on test and the hood are then filled with the explosive mixture and detonated with a spark plug. If transmission occurs, the gas within the hood explodes and the film becomes disrupted. The test is then, if necessary, repeated in the dark without the hood to see the actual point of exit of the flame. The jet of flame can be observed through an armoured glass window in the side of the test cell.

Before the test, the correct gas/air mixture is checked at a sampling point in the exhaust side of the gas circuit. A typical arrangement is shown in Figure 6.3.

6.2 Permitted flameproof gaps

In addition to the test requirements, the flameproof standard specifies maximum clearances and gaps which may be used. They are not relevant for test purposes but are a guide for the designer of the product.

Safe apertures for all commercially used flammable gases and vapours have been determined using standardised test apparatus such as shown in Figure 6.4. The results are termed the maximum experimental safe gap (MESG) for the gas concerned. More recently, theoretical work by Dr.H.Phillips of HSE, Buxton, has enabled the MESG for new flammable gases to be predicted.

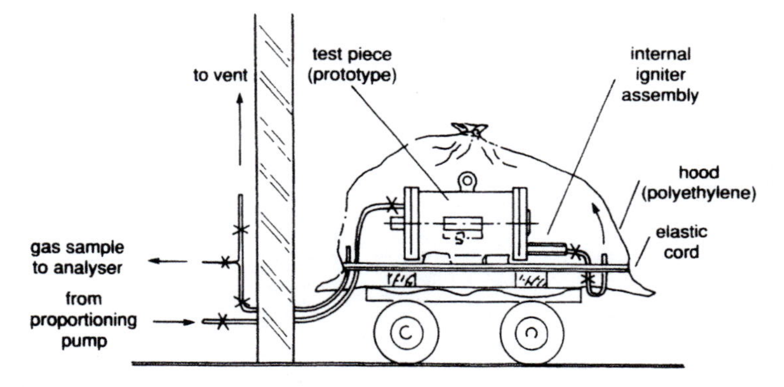

Figure 6.3 Typical method of flame transmission testing

The apertures allowed in flameproof enclosures are those formed by the tolerances and running clearances between separate components. No open holes in a single component, however small, are permitted and all unused entries must be suitably plugged or sealed. The geometry of the gaps can be flat, discoid, cylindrical or compound, as indicated in Figure 6.5 and the respective permitted spacings between the adjacent surfaces are closely specified in the national and international flameproof standards. Experimental results have shown that the maximum values to prevent external ignition vary according to the following parameters

- The internal free volume of the enclosure
- The length of the flame path through the gap

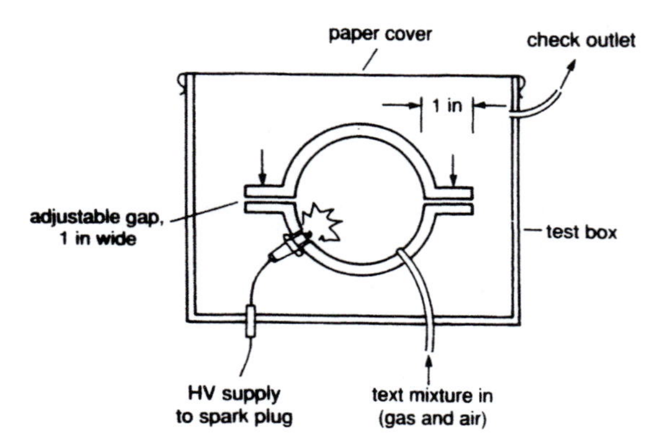

Figure 6.4 Method for the determination of the maximum experimental safe gap (MESG) for a flammable gas using a standard spherical enclosure with an adjustable flange gap

The internal and surrounding spaces are filled with the specified mixture of gas and air and the internal space is ignited. The test is repeated with a successively reduced flange gap until the outer space fails to ignite. This is then the MESG for the gas

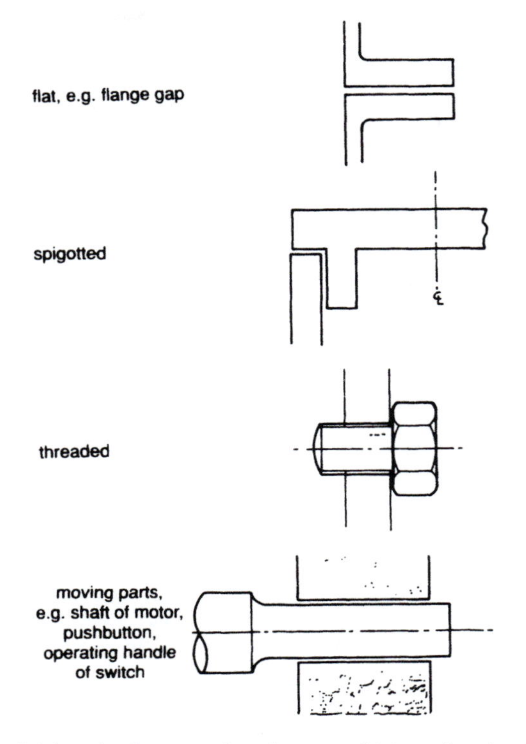

Figure 6.5 Typical joints in flameproof enclosures which need to be closely specified and prototype-tested. External holes should be sealed with eight engaged threads for 'all gases' certification

- The type of gap, e.g. flange, cylindrical, spigot (as indicated in Figure 6.5)
- The flammable gases in which the enclosure is intended to be safe.

In the EN standard, each of these parameters is grouped into a maximum of four ranges for simplification, for instance the enclosure sizes are grouped into volumes <100 cm^3, 100–500 cm^3, 500–2000 cm^3 and >2000 cm^3. Even so, there are some 200 dimensional values incorporated in the standard and 11 different values for the maximum gap. The running clearances.are usually referred to as 'flame paths' and the flame tight joints are often termed 'flange gaps', even though in practice they are totally closed.

For any gap to be effective in preventing an external ignition it must have dimensions below certain critical values and each explosive gas has different characteristics as regards safe flame paths. It is not always economic to manufacture electrical enclosures to the precise tolerances needed to be safe in all explosive gases and vapours. Consequently, flameproof equipment has been classified into four levels of safety, commonly known as 'Gas Groups'. Some of the labelling for these is shown in Table 6.1.

Enclosures with the coarsest apertures are safe only in methane (i.e. firedamp, marsh gas and North Sea gas), methane CH_4 being the most difficult of the dangerously explosive gases to ignite. This is called Group I apparatus and is used only in mining where the gas hazard is always due to methane released from the coal formation. Non-mining equipment is termed Group II apparatus and is further classified into three sub-groups; IIA, IIB and IIC. It should be noted here that American designs and European designs to earlier (national) standards will be classified and marked somewhat differently; see Table 6.1.

IIA apparatus has flame paths which are safe in petroleum type gases and vapours. IIB apparatus has smaller and longer flame paths which are safe in both petroleum and most of the more sensitive ethylene type gases. IIC apparatus has the tightest flame paths of all and is designed to be safe in any explosive gas including hydrogen the most sensitive. There are some abnormal gases, such as acetylene, where the safe dimensions given in the standard may be unreliable. Special gas testing may then be required. The problem with acetylene C_2H_2 is that, with rich explosive mixtures, burning particles of soot can be ejected through the gaps. This is commonly avoided by the adoption of angled flame paths, or spigots.

It has been quite widely believed that the maximum gaps specified in the standards have to be preserved by means of shims etc. to act as a pressure relief in the case of internal explosion. This is quite incorrect and has no doubt persisted because the original British flameproof standard permitted the incorporation of a pressure relief device in the wall of the enclosure. This is no longer allowed and the correct working condition is for all cover bolts to be hardened up and with no gaps between machined flanges. Another false assumption is that the important effect of the flame path is to cool the burning gases before reaching the surrounding atmosphere. This is not the case since a gap between plastic flanges

Table 6.1 Equipment marking indicating range of explosive gases in which flameproof enclosures may be used

	Certification standard				
Degree of flameproofing	Germany (VDE)* Japan (JIS)	UK (BS 229)*	UK (BS 4683 Part 2)*	CENELEC (EN 50018)	USA (NEC)
All gases	Ex d 3n	—	Ex d IIC	EEX d IIC	Class 1 Group DCBA
All gases except hydrogen	—	Group III	Ex d IIB	EEx d IIB	Class 1 Group DCA
Approx. 90% of all known flammable gases§	Ex d 1	Group II	Ex d IIA	EEx d IIA	Class 1 Group D
Acetylene	Ex d 3c	—	—	EEx d IIC	Class 1 Group A
Methane (mining equipment)	Sch d	Group I	Ex d I	EEx d I	—

*These standards have been superseded by EN 50018, but some equipment with the old marking will remain in existence—probably indefinitely.

§The excluded gases are listed in the respective certification standards and documents, e.g. EN 50014: 'General requirements for electrical apparatus for potentially explosive atmospheres'.

can be equally effective. The main cooling of the flame actually takes place outside the enclosure where it becomes entrained with the cooler unignited mixture. It is analogous to trying to light a high speed gas jet with a match, but in this case the flame is moving and the gas is stationary.

6.3 Methods of connecting flameproof equipment

Because the integrity of a flameproof enclosure is paramount as regards its safe use in flammable atmospheres, it has to be designed and used in a way which is reasonably tamper-proof. The installation and maintenance of the equipment is thus a specialised task which is not normally covered by general wiring regulations and electricians should have special training for such work.

The way in which the apparatus is connected, whether by cable or by conduit, is important in preventing external ignition. Methods vary widely from one country to another since the national flameproof standards have treated the problem in accordance with their respective wiring practices. The main alternative means of connection are shown in Figure 6.6. EN 50018 has had to permit all of these solutions and hence does not reduce variety in the field. American installations

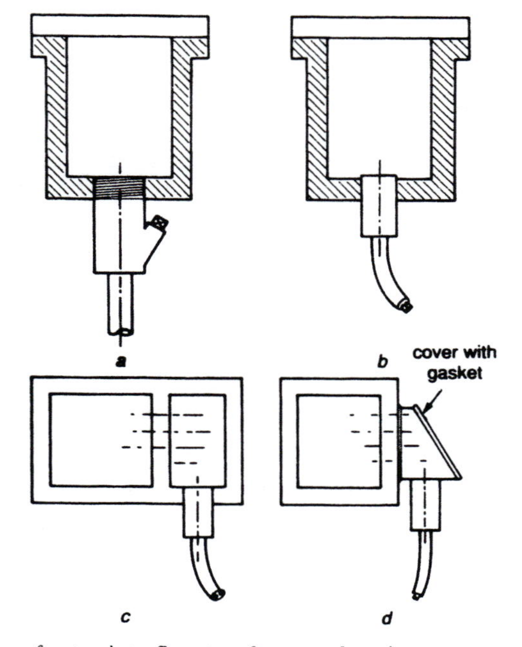

Figure 6.6 Types of entry into flameproof-protected equipment

a direct conduit entry with conduit sealing box
b direct cable entry with gas tight gland
c indirect entry via flameproof terminal box with flameproof bushings to main enclosure
d indirect entry via an increased safety terminal box with flameproof bushings through the side of the flameproof enclosure

remain incompatible with any of the European systems because, in US Division 1 (= Zones 1 & 0) areas of hazard, NEC (NFPA) rules require all flameproof entries to be made through conduit type fittings—which have taper threads. Conversely, in Germany, due to bad experience with damp and corroded conduit, the chemical industry there insists on cable entries for equipment in flammable atmospheres. Furthermore, German practice does not permit cables to enter directly into a flameproof enclosure except via an increased safety (Ex e) terminal box. Thus, a flameproof motor would have an Ex e terminal box with flameproof bushings between the terminal box and the motor frame.

The practice of having an increased safety terminal enclosure fixed to or surrounding the flameproof casing is used both for reasons of safety and of convenience. Although flameproof housings are certified as safe when they produce an internal gas ignition, they are not required to withstand short-circuits or sustained arcing. Under these conditions it has been found that many flameproof designs are capable of causing an external explosion by the bursting or burning through of the casing, even though they comply with the standard and have been duly certified. The adoption of increased safety terminal enclosures, with flameproof protection for the normally sparking components, therefore, has much to commend it on safety grounds.

It also provides simpler installations for the following reasons:

- The terminal box does not need to be a pressure vessel
- The gland plate or enclosure wall at the point of entry does not need to be of the 5 mm or 8 mm specified thickness as required for flameproof gland plates and the entry of cable glands, and conduit can be through plain, non-threaded holes
- All coverplates can be fitted with weatherproof gaskets without the need for spigots, threads or machined faces, as with flameproof covers
- Tamper-proof fasteners are not required
- The same terminal box specification is suitable for all the flameproof sub-groups
- Conduit entries do not need to be fitted with stopping boxes
- Simple glands in place of flameproof types can be used
- Specially filled cable is not required.

In the earlier British standards, direct entry into flameproof enclosures containing normally sparking equipment such as switchgear and motors with slip-rings or commutators was similarly prohibited, but in this case the entry had to be through a subsidiary flameproof terminal enclosure—a practice still commonly used in coal mines in the UK. Effectively, this means that all apparatus except terminal boxes have to be built in the form of two separate flameproof enclosures with factory-made flameproof bushings through the common partition wall. The main enclosure contains the primary apparatus and the second enclosure is drilled and tapped for the user's cable gland or conduit. The ends of the bushings will normally form the connection terminals.

Reference has been made to the use of filled cables. There is no definition given for the effective filling of cables but American practice calls for cables into 'explosion-proof' (i.e. flameproof) enclosures to be sealed round the individual cores if the cable can pass more than 0·2 litres/hour with a pressure head of 15 mbar. If not sealed, the cable must be long enough to have the same effect. However, this rule takes no account of the possibility of precompression* through long cables with an open structure. Intrinsically flameproof cables have never been defined although it is clear that the mineral-insulated metal-clad types are superior in this respect.

All the work and knowledge put into the design, manufacture and testing of flameproof equipment can be readily invalidated by incorrect installation. The cabling and wiring is probably the most important aspect here since the testing of the equipment by the certification authority is generally undertaken without specified cabling into the enclosure. Many of the industrial nations have developed quite different codes of practice for connecting into flameproof housings and, despite attempts at harmonisation for over half a century, this area still causes technical barriers to international trade.

Safe use also requires a knowledge of the field of application for which the apparatus was designed and certified. Misunderstanding the labelling can thus lead to dangerous conditions. Marking codes used by American manufacturers are different from the harmonised CENELEC codes and older designs still constructed to pre-CENELEC standards will have various markings according to the country of origin. There are many documented examples of apparatus which has caused severe explosions due to misapplication.

Before commissioning flameproof equipment all unused entry holes for cables etc. must be sealed and all cover fasteners suitably tightened. If left loose, an internal explosion may be able to produce an unsafe gap. Overtightening must also be avoided. Where cover bolts screwed into cast iron or aluminium enclosures are unduly prestressed, the additional force imposed by an internal explosion may strip the threads. Recommended torque values would be useful for both the tester and the user.

6.4 Applications and limitations of flameproof protection

Protection by flameproof enclosure assumes that ignition of an explosive mixture within the apparatus may occur but that this will not ignite a surrounding explosive atmosphere. This type of protection is therefore suitable for all types of apparatus, including switches, contactors, commutator and slip-ring motors, and incandescent lamps. *Prima facie*, apart from practical limitations of size when the cost of the explosion-resistant enclosure becomes uneconomic, a flameproof enclosure can enable any electrical item to be used in an area subject to the occasional presence of flammable gases and vapours. Its inherent robustness is also of significant benefit in many industrial situations. Examples of typical flameproof electrical equipment are shown in Figure 6.7. There are, however, a number of limitations in the conditions under which flameproof enclosures should be used. The more important of these are described below.

6.4.1 Proximity effects and weatherproofing

We have noted that the main reason for a gap being 'flameproof' is the rapid cooling of the hot flame after ejection, by turbulent mixing with the surrounding

*Also known as pressure piling

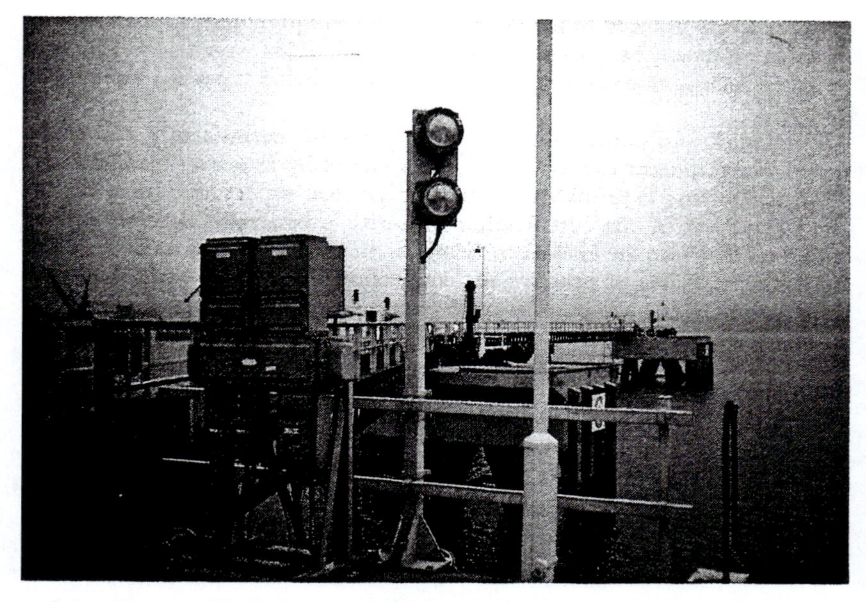

Figure 6.7 Flameproof switchgear and lighting equipment on a tanker-loading jetty

atmosphere. If the ejected flame is slowed down at the point of exit, the MESG is found to be reduced. Accordingly there are some safety restrictions—particularly with Group IIC equipment—as regards weatherproofing by taping flameproof joints and mounting apparatus with its flame gaps close against another surface, either of which can impede the free dispersion of the flame or hot gases.

6.4.2 Battery room protection

These areas are often equipped with flameproof apparatus certified for use in an area containing hydrogen in the most explosive proportions with air. In principle this is not safe because a hydraulic battery under abnormal conditions will produce pure hydrogen and oxygen in the most explosive proportions; a stoichiometric mixture, as formed by the electrolysis of water. This is not covered by any flameproof standard and the effectiveness of certified flameproof equipment in such an atmosphere will not have been proven by the flameproof tests.

6.4.3 Areas above normal atmospheric pressure

Flameproof equipment is not suitable for areas with an over-pressure (such as in tunnelling shields) since the internal explosion pressure will be proportionately increased. If used at twice the barometric pressure for instance, the explosion pressure will also be doubled and may exceed the safety factor provided by the certification over-pressure test.

6.4.4 Low temperature applications

The use of flameproof enclosures at very low temperatures requires special care as explosion pressures can be higher under some cryogenic conditions. For example, Underwriters Laboratories in America have recorded that a 250 hp motor filled with ethane and air at 22°C produced an explosion pressure of 65 lb/in^2. When the initial temperature was reduced to – 70°C, the subsequent peak explosion pressure was 250 lb/in^2. This situation is of course aggravated by the embrittlement of the casing at that temperature.

6.4.5 Burning insulation and particle ignition

In the 1974 annual report of the Safety in Mines Research Establishment (SMRE) it was pointed out that during the previous few years several flameproof enclosures had burst due to the burning of organic insulation within the enclosure. This was found to be due to the very rapid generation of gases (mainly hydrogen) by pyrolysis of the plastics under sustained arcing conditions. The SMRE experiments also demonstrated that, even when the enclosure remained intact, the gas could be ignited outside the casing by incandescent particles from the arcing electrodes which were expelled along with the gas.

The worst materials in general use in respect of gas evolution are the phenol formaldehyde resins. HSE has since carried out work in this field enabling a safer form of flameproof apparatus to be designed. This work has yet to be reflected in any of the certification standards. So-called particle-proof joints with angled flame paths have been suggested.

6.4.6 Electrical fault levels within the enclosure

An electrical short-circuit or earth fault is one possible cause of an internal gas ignition. The type testing by the certifying agency will have ensured that the housing will withstand the force of the gas explosion. In the over-pressure test the gas mixture is exploded by a low power igniter, such as a spark plug. This gives no assurance that the casing will also withstand higher pressures due to electrical energy released in the fault itself. A gas explosion in a closed vessel will create a peak pressure of up to about 30 atmospheres, but an electrical fault generates energy which can produce far higher pressures.

The combustion energy released per litre of explosive gas mixture is about 0·2 kJ whereas the energy liberated by a short-circuit on a system with a fault level of 20 MVA for example may be up to 200 kJ per cycle. Clearly it would be preferable if flameproof enclosures were designed to be 'fault retaining' to a specified level of let-through energy, but here again this requirement is not included in the flameproof standards accepted either by CENELEC or IEC.

6.4.7 Persistent arcing within an enclosure

As long ago as 1944, the then Ministry of Fuel and Power pointed out in an HMSO publication that a fault within a flameproof enclosure may result in arcing severe enough to burn a hole through the casing and that suitable automatic electrical protection for the circuit is necessary to guard against damage of this description.

No flameproof standard has yet addressed this possibility, although North American explosionproof standards specify a minimum thickness for aluminium enclosures.

6.5 Review

In conclusion, we can say that flameproof protection is one of the oldest forms of design for electrical products used in potentially explosive atmospheres. It is suitable in the majority of cases for all but the largest classes of equipment, where exclusion of flammable gas by internal pressurisation (Ex p) is often more appropriate. Flameproofing is accordingly widely used by both mining and surface industries. There are, however, limitations in its field of application which are not always evident from the standards and codes of installation practice.

The additional cost of the pressure-resistant housing and other restrictions make this type of protection uneconomical for small, low-power apparatus, such as that occurring generally in process control and information technology. For these areas, the type of protection—intrinsic safety (Ex i)—is usually a more suitable choice.

6.6 Bibliography

PHILLIPS, H., *Estimation of the maximum experimental safe gap for a fuel by various methods.* Research Report No.259, Safety in Mines Research Establishment, Buxton, Derbyshire. 1969

BS 5501 Part 5: 1977 [= EN 50018: 1977], *Flameproof enclosure 'd'*, Publ. BSI

BS 5345 Part 3: 1990 *Installation and Maintenance requirements for electrical apparatus with type of protection 'd'. Flameproof enclosure*, Publ. BSI

A review of electrical research and testing with regard to Flame-proof Enclosure and Intrinsic Safety of electrical apparatus and circuits, Publ. HMSO, 1944

Chapter 7

Protection by intrinsic safety

To set off an explosion of a flammable gas three factors are necessary; fuel, i.e. the gas, oxygen, as contained in air and a source of energy, usually in the form of heat, to trigger the explosion. To pull the trigger a minimum amount of energy must be transferred to the gas mixture and below this minimum level an explosion will not take place. The amount of energy required is different for each gas, hydrogen being the most sensitive.

There are two obvious ways in which a flammable gas can be ignited from an electric circuit or apparatus:

(a) By arcs or sparks
(b) By hot wires or components.

With (a), the critical parameters are the energy released in the spark and the ability of the spark to transfer a sufficient amount of this energy to the explosive mixture. The energy liberated in a spark caused by a break in an electric circuit depends upon the current flowing at the time of the break and on the restriking voltage across the sparking electrodes. Sparks, however, are not only caused by breaks due to disconnections but also by accidental contacts causing earth faults and short-circuits. At the moment of contact, some metal will be vaporised so that a spark is momentarily created. The criterion for intrinsically safe circuits is therefore not just the normal current in the circuit but the possible current which could flow under the conditions of a short-circuit.

The electrical parameters needed to cause an igniting spark have been found to vary with a number of other parameters which affect the transfer of energy to the explosive mixture. These include the type of contact (or electrode) metal, and speed of separation of the sparking contacts. A certain period of arcing of a few hundred microseconds is also necessary to give the heat from the spark time to enter the gas.

For ignition by method (b), the critical parameters are the surface temperature, the area of the heated surface exposed to the gas and the relative velocity of the flammable atmosphere over the surface.

Work on intrinsic safety was pioneered in the UK and used, in British coal mines in particular, some 30 years before the technique was generally recognised overseas. A British Standard on the subject has existed since 1945 and in the subsequent 20 years over 500 certificates for intrinsically safe designs were issued by the Electrical Branch of the Factory Inspectorate, based on test reports carried out by the Safety in Mines Research Establishment—both now part of the Health and Safety Executive.

7.1 The design of intrinsically safe systems

The principle of protection by intrinsic safety is to keep both the voltage and current available to the system to such low values that the energy which can be released, even under fault conditions, is too low to be able to ignite a mixture of explosive gas and air.

Research has shown that the lowest current which can cause an explosion falls very rapidly with increasing supply voltage. This is mainly due to the increasing ability of the source to maintain an arc as the voltage is increased. The relationship between the maximum safe open-circuit voltage and the maximum safe short-circuit current is, in fact, roughly in inverse proportion, as indicated in Figure 7.1. If the voltage and current at the spark lie below this curve, the circuit is said to be intrinsically safe, or IS at that point.

Although each gas relates to a different safe curve, they are all of the same form, being more or less 'parallel' and lying between the extremes of hydrogen, the most sensitive and methane, the least sensitive. It is therefore possible to design an IS system for the most sensitive gas concerned and to know that it will not ignite

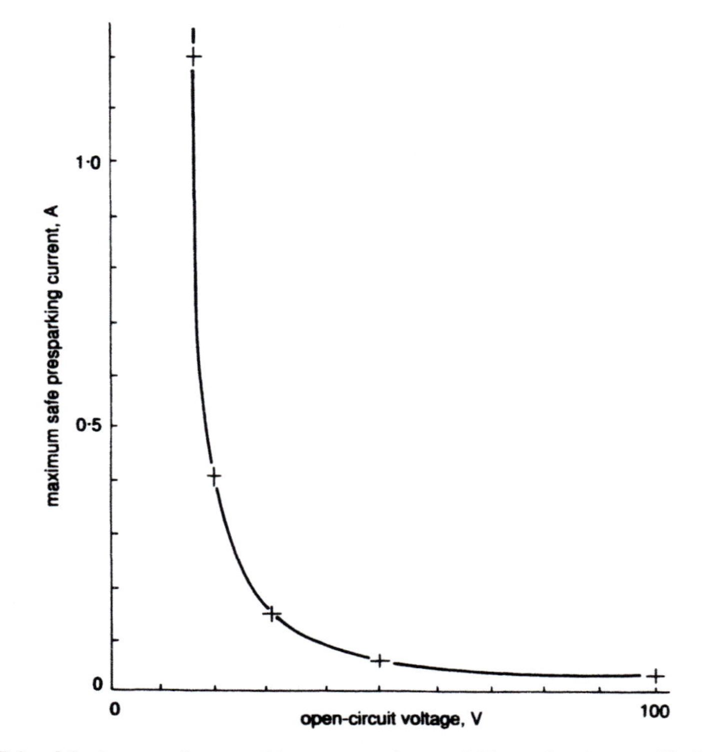

Figure 7.1 Maximum safe presparking currents for a stoichiometric mixture of hydrogen and air

any less sensitive gas. In other words, none of the curves intersects with another. For commercial reasons, four standardised curves are used. These have been selected so that they respectively cover virtually the same groups of gases as the four flameproof groups. This enables a uniform system of labelling to be used for both flameproof and IS equipment. Thus, IS apparatus certified to the harmonised European standard for intrinsic safety (Ex i) protection will be marked:

EEx i IIC Safe in all flammable gases
EEx i IIB Safe in all flammable gases except hydrogen
EEx i IIA Safe only in the IIA-listed gases in the general standard EN 50014
EEx i I Safe in methane, for mining use.

It will be seen from the curve in Figure 7.1 that at the higher voltages very low values of current in the spark can cause ignition, whereas in the region of 12 V, there is apparently no current capable of causing ignition. This is of course not the whole story. A spanner dropped across a car battery, for instance, would certainly be able to ignite petrol vapour, but in this case the transfer of heat to the gas would be from incandescent particles of metal rather than from the actual spark. Further examination of the curve shows that the allowable power in the circuit increases markedly at the lower voltages. Thus, with a 100 V source, a 25 mA spark is the safe limit, corresponding to a circuit power of 2·5 W. With a 15 V source, 1·2 A is permissible, corresponding to a circuit power of 18 W. Intrinsic safety is accordingly a method of protection against ignition which lends itself very well to instrumentation circuits and solid state technology, where very low voltages and relatively low levels of power can be used. Figure 7.2 shows the maximum safe power which can be allowed in a spark at various voltages.

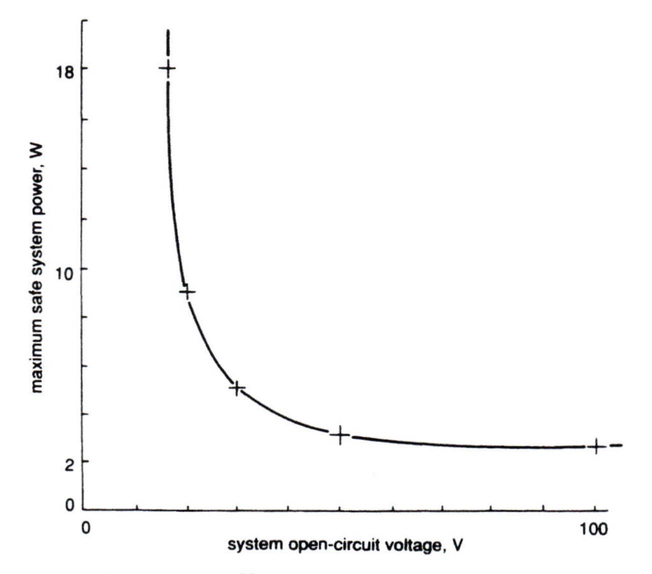

Figure 7.2 Increase in permissible IS power source as system voltage is reduced (useful power at matched load impedance will be half these values)

7.2 Applications of IS systems

Because IS protection is clearly more applicable to low power and particularly to very low voltage equipment, it has been widely adopted for the explosion protection of electronic process control and telemetering systems where all or part of the circuit has to be within a hazardous area. A hazardous area in this context means a zone where an explosive atmosphere could be present, a tank farm or pumping area for example. Other examples are shown in Figures 7.3 and 7.4.

Low voltage solid state technology and IS techniques have advanced rapidly together, so that nowadays more electrical goods are being designed and certified for intrinsic safety than for any other type of explosion protection.

Most of the apparatus and circuit wiring forming an IS system are normally installed and operated in a safe area such as a control room. Those parts of the IS system which have to be in a hazardous area are usually restricted to simple probes and devices such as float switches, thermocouples, transducers for measuring and transmitting process conditions and current-to-air pressure transformers for servocontrol. These items present no hazard in themselves as they cannot generate any energy to ignite gas/air mixtures. No adjustments carried out on them can make them dangerous and often they do not even need to be tested for intrinsic safety.

Figure 7.3 North Sea gas well drawgear—an IS instrumentation zone

Figure 7.4 North Sea gas reception area — an IS instrumentation zone

However, the equipment in the safe area, to which they may be connected, can create a very considerable hazard by allowing dangerous levels of energy to pass from the safe area to the hazardous area loop. It is necessary in most cases to assume that a mains voltage above earth will accidentally become connected at some time to the IS system and so produce a dangerous energy source in the hazardous area. This is the Trojan horse syndrome of IS instruments and probes in hazardous areas. By international agreement, in CENELEC and IEC, the foreseen fault conditions in the safe area will generally include open-circuits, short-circuits, component failure and the unintentional connection of the LV mains (in the UK 230 V a.c.) to some part of the IS circuit.

Hence, most of the recent development of IS systems has been concerned, not with the hazardous area apparatus at all, but with the safe area circuits and devices which may be connected to it. In fact, IS protection has very little to do with apparatus in the hazardous area, but is almost entirely concerned with what the hazardous area items are connected to in the safe area, for, in general, this is where the dangerous sources of energy will lie.

The problems of ensuring intrinsic safety can be greatly simplified by installing suitable power transfer limiters at the point where the loop into the hazardous zone is connected to the equipment in the safe area. These so-called barrier devices can take the form of optocouplers, saturable transformers, fused Zener diode circuits and even the simple electro-magnetic relay. The choice will depend primarily on the type of information (analogue or digital) and required rate of data to be dispatched or transferred and the power to be transmitted.

As shown diagrammatically in Figure 7.5, barrier devices should be installed in the safe area, as near as possible to the hazardous area they are protecting. This is because the barrier can only protect its downstream circuits from spurious voltages and energy sources which enter upstream of the barrier. It cannot protect against dangerous voltages entering the circuits beyond the barrier. By using certified barrier devices at the boundary between the safe and hazardous zones, the equipment and circuit wiring in the safe area—such as VDUs and process

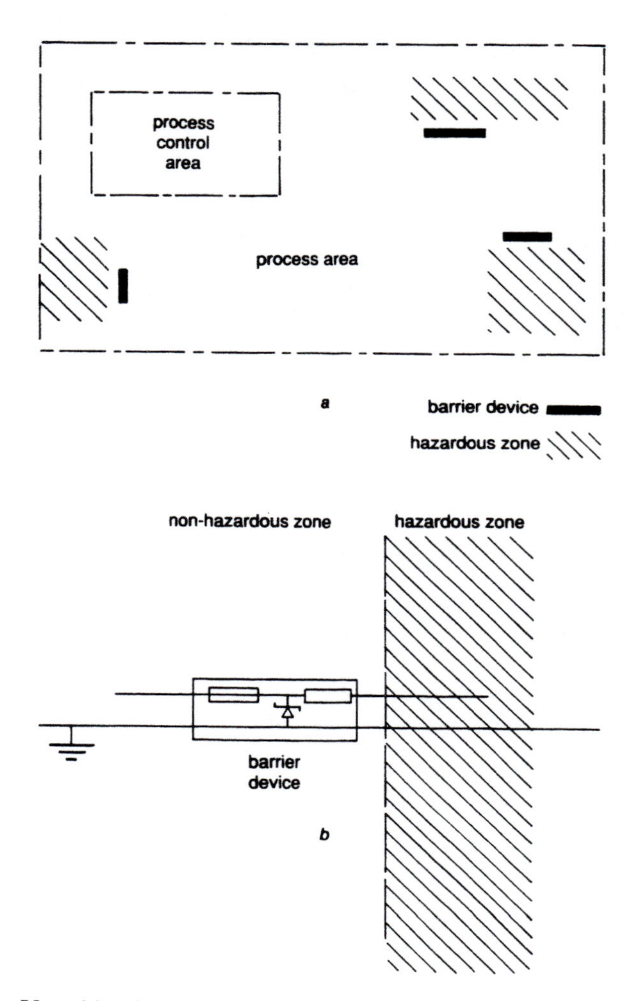

Figure 7.5 Use of barrier devices to limit fault power invading a hazardous area via an IS system

a Optimum position of barrier devices with respect to hazardous zones
b Principle of a commonly used type of barrier. The current-limiting resistor enables the device to be connected to a low source impedance. The fuse protects the Zener diode from undetected damage due to excessive avalanche current

controllers—do not have to comply with the stringent reliability requirements of the IS standards. Danger can arise either due to a voltage being imposed which is too high or because the source impedance of the power supply to the system becomes too low. Because of the barrier device, any unsafe voltage or unsafe source impedance will not reach the IS loop in the hazardous area.

7.3 Testing for intrinsic safety

As stated above, for safety reasons faults have to be taken into account in all parts of the circuit, but more particularly in those parts which extend into the safe area where more carefree methods of maintenance and adjustment may occur. It is also in the safe area where the IS system will usually be connected to its source of supply. Laboratory testing for IS certification therefore involves an assessment of the whole circuit or system in both hazardous and safe areas. If certified barriers are not to be used, possible faults in the safe area must be applied or simulated and at the same time the hazardous area part of the circuit is interrupted and the resulting spark tested for incendivity in the gas/air mixture concerned. A sparking test device filled with the specified explosive gas mixture is used for this purpose.

At this point it is necessary for us to consider inductive effects. To produce a spark from an electric circuit the flow of current must be interrupted, or the voltage must be increased to a value above the breakdown level of the dielectric. In the case of IS systems we are concerned only with low voltage and very low voltage circuits so that dielectric breakdowns and flash-overs can be disregarded. When a low voltage circuit is broken a spark and a restriking voltage is produced. The potential difference or volt-drop across the spark gap is determined by the source voltage and the rate of collapse of the magnetic flux. Consequently, since the energy in the spark is the critical factor, the greater the inductance in the circuit, the lower the current which will be able to ignite gas when the circuit is broken. High inductances can, of course, ignite more than explosive atmospheres as the energy in the magnetic field can be very considerable (sub-station engineers used to light their cigarettes in the days of live-front panels by simply slightly opening a standby generator field switch).

Taking circuit inductance into account, the limiting curve for intrinsic safety shown in Figure 7.1 becomes modified to the form shown in Figure 7.6. Here, the safe current level continues to increase as the voltage is reduced, until a value of current is reached at which the restriking voltage is determined by the circuit inductance rather than by the system open-circuit voltage. The safe current will then no longer be increased by any further reduction of the source voltage. In principle, the maximum safe current should decrease inversely as the square root of the inductance so that the energy $e = \frac{1}{2}LI^2$ remains constant. This can be demonstrated by tests with a sparking device in an explosive gas mixture.

The actual value of the restriking voltage will also increase according to the rate of separation of the sparking contacts. And if the electrodes are moving apart quickly their heat sink or cooling effect on the spark itself will be reduced so that the spark energy will be more effectively transferred to the explosive atmosphere. For these two reasons, the most incendive sparks are produced by sudden breaks in small electrodes—such as by thin wires being broken under tension.

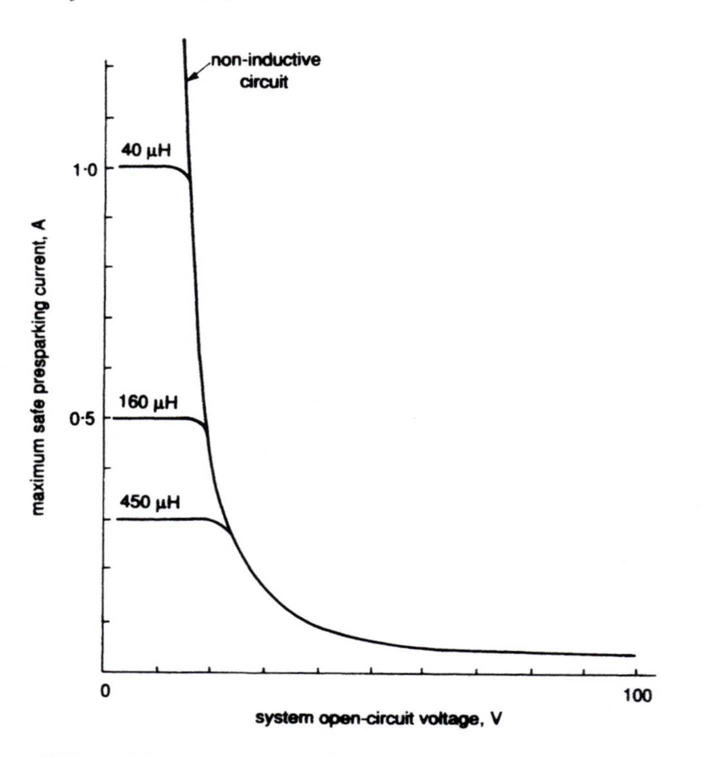

Figure 7.6 *Effects of inductance in an IS circuit. The maximum safe current is reduced approximately inversely as the square root of the inductance (i.e.* $\frac{1}{2}LI^2$ *remains substantially constant)*

It has also been found that some electrode surfaces can ignite gases more readily than others. For instance, as shown in Figure 7.7, a copper wire would not ignite hydrogen by a 50 V spark below 450 mA, whereas if it were cadmium-plated it could cause ignition down to 60 mA. The design of the sparking device used to test circuits for intrinsic safety thus has a profound effect on the test results obtained.

The choice of the test device able to create the most sensitive conditions for ignition has been discussed for over half a century and many different types have been used. The present test apparatus, standardised by the IEC and CENELEC, is based on German test practice and consists of tungsten wires freely scraping over a grooved cadmium disc. It is described in detail in the European harmonised standard for protection by intrinsic safety, EN 50020. The design is intended to produce a series of random sparks since these are found to be more likely to ignite an explosive atmosphere than a given sequence of identical sparks. Figure 7.8 shows the device without its transparent cover which contains the explosive mixture.

Normal test practice for intrinsic safety certification is to vary the circuit conditions using, for instance, an adjustable source impedance or current-limiting resistance (CLR) until a borderline between ignition and non-ignition is found.

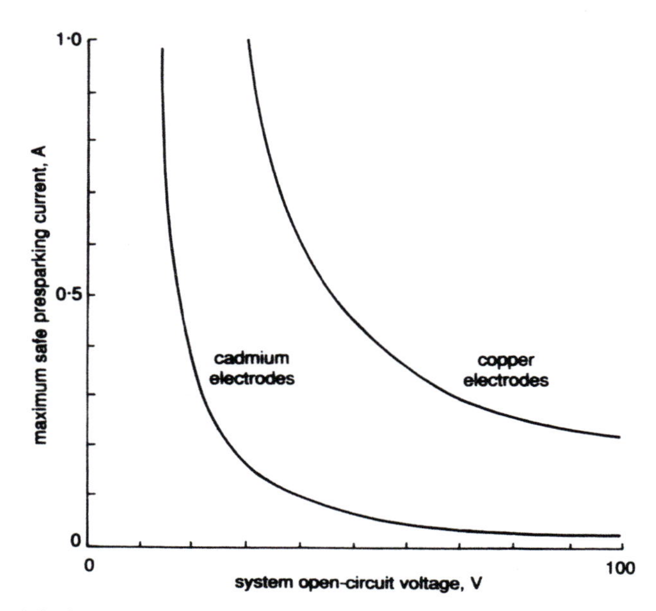

Figure 7.7 *Maximum safe presparking current values for H_2 with copper and with cadmium electrodes*

Figure 7.8 *Standard sparking test apparatus with explosive atmosphere-retaining cover removed. The discs rotate at different speeds, four tungsten wires being attached to the upper disc. The lower disc is slotted and made of cadmium*

A significant safety factor then has to be applied to the result, as the poor repeatability of the test demands fairly wide confidence limits.

Circuits which have a known value of source voltage, source impedance and of stored energy can be assessed for IS by evaluation against the limiting safe curves which are published in the above-mentioned standard. The energy stored in capacitors can be readily determined from $e = \frac{1}{2} CV^2$, but the effective inductance in a circuit can present a problem. The inductance of components containing iron, such as miniature solenoid valves, magnetic relays and chokes, will depend on the slope of the B/H curve. Because this is not a straight line, the inductance in the circuit changes with the degree of saturation and consequently depends on the current value at the moment the circuit is interrupted. Owing to the difficulty of calculating the effective overall inductance in such a circuit, it may then be necessary to demonstrate whether or not it is intrinsically safe by means of the standardised spark test apparatus. It is, in any case, essential to run the spark test apparatus for an extended period in order to ensure that some sparks occur at those instants which create maximum release of energy into the gas.

Restriking voltages in the hazardous area can usually be reduced to a safe level by a suitable arrangement of diodes across each inductive component. As regards the stored energy in capacitors, it may be found desirable to connect a CLR in series to limit the possible short-circuit current to an IS value at the working voltage of the capacitor.

7.4 Ignition by overheated components

As previously mentioned, explosive atmospheres can be ignited by unduly hot wires or components. This imposes an additional restriction on IS systems which is not necessarily covered by the sparking test criterion. Circumstances that have to be considered include wires containing fine stranded conductors, compound resistors and the carbonisation of insulation.

The fraying of a flexible lead down to the last unbroken filament could cause an explosion, because very fine wires can become incandescent even at powers too low to cause spark ignition. For this reason, specific rules concerning minimum filament size in stranded wires are applied.

Low wattage resistors having very small surface areas can also be a hazard as they will readily reach excessive temperatures if overloaded. A 2·5 W resistor, for instance, will generate an increase in its surface temperature of up to 100 K per watt of heat dissipation. If compound-based resistors are overloaded by relatively minor amounts they can char, or carbonise. Their resistance then decreases with temperature so that their loading increases until it equals the source impedance. The component can then remain glowing and burning in this stable matched-load condition.

This type of resistor is accordingly not suitable in critical parts of an IS circuit, that is to say, either in the hazardous area or as a CLR in the safe area. It is preferable to use a single-layer wire-wound resistor on a ceramic base, as this is unlikely to fail in anything but an open-circuit mode.

Carbonisation of insulation can produce a similar effect to the charring of compound resistors and applies in particular to IC devices mounted on organic substrata. Again, a ceramic or glass base is preferable.

For the above reasons, testing therefore also involves some assessment of the possible surface temperatures of components in the hazardous zone. Figure 7.9 indicates the use of a radiation pyrometer to measure the surface temperature of very small PCB components where the attachment of a thermocouple would act as a heat sink and give unduly low readings.

In conclusion, it will be seen that protection by the method of intrinsic safety has grown rapidly due to its eminent suitability for very low voltage, low power process control and information transfer installations. It differs from other types of protection for electrical apparatus in potentially explosive atmospheres because the flow of energy from the safe area must be carefully controlled. Its growth has been greatly accelerated by the development of barrier devices which protect the hazardous area from the possible intrusion of dangerous sources of energy from the safe area parts of the system.

The use of fibre-optics for the transmission of data will no doubt increasingly supersede many existing barrier applications for IS systems.

7.5 Bibliography

REDDING, R.J. and TOWLE, L.C., *Barrier method of ensuring the safety of electrical circuits in explosive atmospheres*, Proc. IEE, Vol.113. No.12, Dec. 1966, pp. 2070-2074

WIDGINTON, D.W., *A method for assessing the effective inductance of components used in intrinsically safe circuits*, Research Report No.254, Safety in Mines Research Establishment, Buxton, Derbyshire. 1968

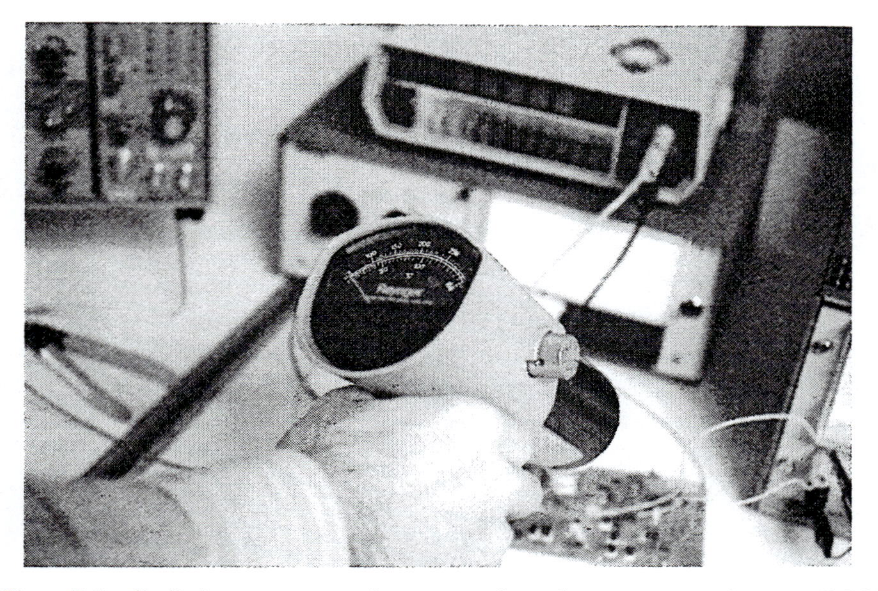

Figure 7.9 Radiation pyrometer used to measure the surface temperature of very small PCB components, where the attachment of a thermocouple would act as a heat sink and give unduly low readings

WIDGINTON, D.W., *Some aspects of the design of intrinsically safe circuits*, Research Report No.256, Safety in Mines Research Establishment, Buxton, Derbyshire. 1968
MURPHY, R.H., *The zener diode—an accurate voltage reference source*, Electronics and Power, Dec.1966, pp. 430 et seq., Publ. IEE
REDDING, R.J., *Intrinsic safety*, Publ. McGraw Hill, London. 1971
Intrinsically safe apparatus for use in Division I hazardous locations 1978, NFPA Standard 493, Publ. National Fire Protection Association, Massachusetts
BS 5501 Part 7: 1977 *Intrinsic safety 'i'*, Publ. BSI*
BS 5345 Part 4: 1977 *Installation and maintenance requirements for electrical apparatus with type of protection 'i'. Intrinsically safe electrical apparatus and systems*, Publ. BSI
BS 5501 Part 9: 1982 *Specification for intrinsically safe electrical systems 'i'*, Publ. BSI#

* BS 5501 Part 7 = EN 50020
\# BS 5501 Part 9 = EN 50039

Chapter 8

Electrical apparatus in areas subject to flammable dusts

8.1 Dust clouds

When a mass of solid flammable material is heated it burns away relatively slowly, layer by layer, owing to the limited surface area exposed to the oxygen of the air. The heat from the combustion is liberated gradually and harmlessly because it is dissipitated as quickly as it is released.

The result is quite different if the same material is ground to a fine powder and intimately mixed with air in the form of a dust cloud. In these conditions, the surface area exposed to the air is very great and, if ignition now occurs, the whole of the material will burn at once. The combustion energy is then suddenly released producing large quantities of heat. The resulting rise in pressure is comparable to that of a gas explosion and equally destructive. The peak pressure can rise to over seven atmospheres in about one tenth of a second. Some dusts will volatilise before burning and explode more like a gas, with the additional possibility of pressure piling effects.

Almost all combustible materials can form an explosive dust cloud in certain circumstances. Unless, therefore, there is positive knowledge to the contrary, it should be assumed that any organic or carbonaceous material may give rise to dangerous dust. This includes many plastics and most natural products of animal and vegetable origin and virtually any foods and fodder in powdered form – not only grain and pulses, but also meat and fish meal, blood, soup powders and even aspirin. In addition to these, many other oxidisable substances will form explosive dust clouds. Of these, metal powders such as iron and steel and particularly aluminium have caused serious problems in the prevention or limitation of explosion damage. Among non-metals, sulphur has produced the greatest number of explosions.

Dust explosions are a world-wide phenomenon. In the USA, for example, between 1958 and 1977 there were about 220 recorded severe dust explosions due to foodstuffs and animal feed alone, in which 148 people were killed and 499 injured. Table 8.1 lists some of the substances which have been involved in major dust explosions. This list represents only a small percentage of the total number of substances which have been the subject of dust explosions in the UK.

Table 8.1 Substances which have given rise to major industrial dust explosions

	Ignition temperature of dust cloud (°C) from Ref. 1	from Ref. 2
Aluminium	590	640
Coal	595 – 655	610
Dextrin	400 – 430	400
Hard rubber	360	350
Magnesium	470	520
Sulphur	235	190

Ref. 1: Ditgens, K. and Hagen, A. Verordnung über elektrische Anlagen in explosionsgefährdeten Räumen, Carl Heymanns Verlag, 1964
Ref. 2: Marks, L.S. Mechanical Engineers' Handbook, 6th Edition, McGraw-Hill, 1968

As can be seen in the table, the ignition temperature, as determined by test, differs somewhat according to the method of test and the shape and size of the particles. It will also be significantly influenced by both the source and pretreatment of the powder. Some differing values are given in BS 7535:1992.

The causes of ignition are often impossible to determine and it has to be assumed that many apparently spontaneous dust cloud explosions arise due to some effect of static electricity (further discussed in a later chapter). Electrical equipment itself is the cause in less than 3% of recorded dust fires and explosions. Accordingly, it is not surprising that most technical papers on this subject concentrate on explosion pressure relief and damage limitation by venting and the prevention of ignition by methods such as inerting and the avoidance of static electricity, rather than on the design of safe electrical equipment.

The lowest concentration of dust capable of exploding when dispersed with air is referred to as the lower explosive limit. This varies with different substances and ranges from some 10 g/m^3 to about 500 g/m^3. Whereas explosive gas mixtures are generally invisible, unless accompanied by a mist, dust clouds above the lower explosive limit will always resemble a very dense fog.

The upper explosive limits are not well defined and are of little importance in practice. The range of explosive concentrations of a dust cloud is to some extent affected by the size and shape of the particles as well as their chemical composition. The finer the dust or powder, the more readily it will be dispersed into a cloud and the longer it will remain in suspension. The ease of ignition, rate of burning, rate of rise of pressure and the maximum pressure produced by the explosion all increase significantly as the average particle diameter is reduced. The critical factor throughout is of course the surface area per unit weight and for this reason, particles in the form of flakes are more readily ignited than spherical particles of the same particle mass.

The US Bureau of Mines uses a hazard rating for dusts based on Pittsburgh coal as a yardstick. The standardised properties of Pittsburgh coal are:

Minimum ignition temperature	610°C	[A_c]
Minimum ignition energy	60 mJ	[B_c]
Minimum dust concentration in air	45 g/m^3	[C_c]

With appropriate units, values of A, B and C give a hazard rating for a dust as regards ignition 'sensitivity'.

The ignition sensitivity for a dust d is defined as

$$\frac{A_d}{A_c} \times \frac{B_d}{B_c} \times \frac{C_d}{C_c} = \text{index X}$$

An index of explosion 'severity' is then calculated from the maximum explosion pressure and the maximum rate of pressure rise, again based on Pittsburgh coal, its standardised properties in this respect being

Maximum explosion pressure: 83 psi (5·8 bar) [D_c]
Maximum rate of pressure rise: 2300 psi/s (0·16 kbar/s) [E_c]

With appropriate units, these values give a hazard rating as regards explosion severity which, for a dust d, is defined as

$$\frac{D_d}{D_c} \times \frac{E_d}{E_c} = \text{index Y}$$

The overall index of 'explosibility' Z is then given by

$$Z = (X) \times (Y)$$

Table 8.2 US Bureau of Mines ratings for some powdered food products

	Sensitivity index	Severity index	Overall index
Cocoa	1·1	1·3	1·4
Skimmed milk	1·6	0·9	1·4
Potato starch	5·1	4·1	20·9
Wheat starch	10·6	4·7	49·8
Sugar	5·5	2·4	13·2
Flour	2·1	1·8	3·8
Coal	1·0	1·0	1·0

Table 8.2 shows that under this system of evaluation the starches present the most serious hazards.

The safety of electrical installations in hazardous atmospheres is governed by the possibility of the apparatus being able to cause an ignition. With explosive gases and vapours, three ignition characteristics need to be considered:

(a) The minimum temperature at which the gas or vapour will spontaneously ignite
(b) The minimum energy required to ignite the gas or vapour (relevant to protection by intrinsic safety)
(c) The minimum gap dimensions capable of transmitting an incendive flame (relevant to protection by flameproof enclosure).

For combustible dust clouds it is practicable to design the enclosures of electrical apparatus to prevent the entry of hazardous quantities of dust. It is then only necessary to consider ignition characteristic (a). Thus, for electrical apparatus for use in areas subject to combustible dust clouds, the sealing and outer surface temperature of the housings are the main safety factors as regards prevention of ignition.

8.2 Deposited dust

Explosive dusts are generally confined to closed vessels such as silos, cyclone extractors, grinding mills etc. However, where layers of dust are allowed to settle in industrial premises it can form a dangerous atmosphere whenever it is disturbed. Coal dust lying in mine roadways presents a particular hazard of this type. In surface industries, if deposits of flammable dust are left to accumulate, they will be similarly disturbed by a gas or dust explosion and so add to the conflagration. Deposited dust may also become ignited *in situ* where it lies on an unduly hot surface. Oxygen will have limited access at the base of the deposit and dust fires of this nature often start with a prolonged period of smouldering below the surface. This initial stage is usually possible at a surface temperature far lower than that required to ignite the same dust in suspension as a cloud. For instance, Ref. 1 of Table 8.1 also states that a 5 mm layer of coal dust will begin to smoulder or glow at a surface temperature of 235°C, whereas a surface temperature of 595 to 655°C is required to ignite a coal dust cloud.

If organic dusts are contaminated with oil, smouldering can start at substantially lower temperatures—even down to 100°C. A paper by P. C. Bowes and B. Langford at the Fire Research Station, Hertfordshire, drew attention to fires starting in oil-soaked lagging round pipes at 80–150°C. Again, the first stage is smouldering and glowing of the lagging and flames may not appear until the outer covering of the lagging is pulled away.

8.3 Surface temperatures

For electrical equipment installed in areas subject to combustible dusts the surface temperature of the enclosure must obviously lie below both the cloud and layer ignition temperatures of the material concerned, which may be in the form of dust, powder, fibres or fluff. In practice it is the smouldering temperature of a deposit which will be critical since this is generally lower than the surface temperature needed to ignite the material in suspension. Some substances, however, cannot readily be ignited as a deposit on a heated surface because the material first melts or carbonises. Sulphur dust is one example.

The surface temperature of pipes and vessels containing hot fluids and gases will rise when they are thermally lagged by a deposit of dust, but it will not go above the temperature of the contents. The situation is quite different with electrical apparatus as in this case there is a source of heat. Consequently, with electrical housings and cables, if the lagging effect due to dust deposits is increased, the surface temperature at the base of the dust layer must rise until the same overall heat loss can flow. The surface temperature to be considered for fire risks should accordingly be that which occurs under these conditions. This means in turn that, where electrical apparatus is to be operated in areas likely to have deposits of dust etc. the maximum thickness of the deposit must be specified.

The differing effects of thermal insulation are shown in Figure 8.1. An oily atmosphere in conjunction with a flammable dust forms a particularly dangerous combination. If the dust is oily or sticky, the cooling ducts of electric motors can become clogged so that the machine runs hotter than it should and its life becomes shortened due to the more rapid ageing of the insulation. Generators driven by diesel engines sometimes suffer in this way on account of the oily atmosphere from the engine. Oily dust adhering to lighting fittings may obscure the light and make them hotter as well as less effective. Also, as mentioned above, organic oils reduce the ignition temperature of many types of combustible dust and flyings.

8.4 Design of apparatus

In general there are only two basic requirements:

(a) The enclosure should be constructed so that flammable dust will not enter the interior
(b) The temperature of the outside surfaces must not be high enough to ignite stirred up dust nor cause the onset of smouldering of deposited dust.

In any case, thick layers of dust should be avoided wherever possible by suitable means, such as

- Designing the enclosure with steeply sloping top surfaces
- Placing a suitable hood over the apparatus
- Installing the apparatus in a clean room

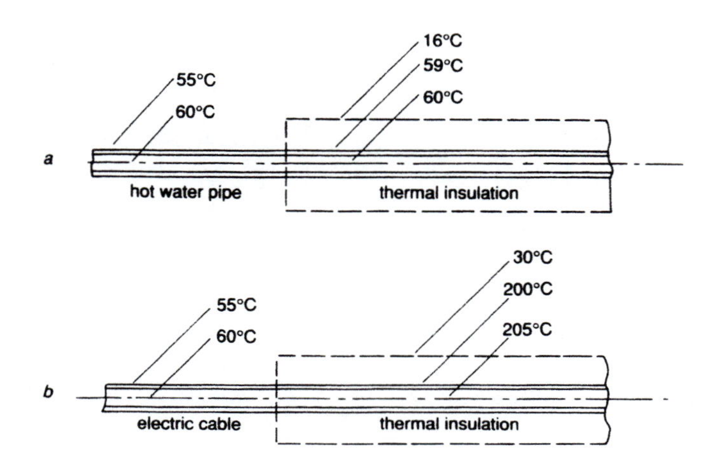

Figure 8.1 *Typical thermal effects caused by thermal insulation such as by deposited dust*

a For a hot water pipe
b For an electric cable carrying full load current

In this example the hot water pipe surface temperature is raised only 4°C by a dust layer, whereas the surface temperature of an electric cable is raised by 145°C

- Suitable dust control (e.g. as seen in Figure 8.2)
- The use of extractor fans
- Regular good housekeeping to remove accumulated dust.

There are three British Standards concerning the design of electrical equipment for use in the presence of combustible dusts:

BS 6467: Part 1: 1985
Electrical apparatus with protection by enclosure for use in the presence of combustible dusts
Part 1. Specification for apparatus

BS 6467: Part 2: 1988
Electrical apparatus with protection by enclosure for use in the presence of combustible dusts
Part 2. Guide to selection, installation and maintenance

BS 7535: 1992
Guide to the use of electrical apparatus complying with BS 5501 or BS 6941 in the presence of combustible dusts

The two standards referred to in the title of BS 7535 address the requirements for electrical equipment in explosive gas atmospheres.

These standards are constructed on the basis of two levels of hazard:

Zone Y Where layers of dust capable of forming an explosive cloud are possible but unlikely
Zone Z Where an explosive dust cloud may be present during normal processing, handling and cleaning operations.

Figure 8.2 Dust control in a carpenter's workshop

Courtesy of North Brink Cabinets, Wisbech

For Zone Y, enclosures complying with grade IP 5X of the ingress protection code based on IEC Publ. 529 are specified. For Zone Z, grade IP 6X is specified.

Enclosures to IP 5X are termed dust protected. They will not totally prevent the ingress of dust but the quantity which will enter is regarded as insufficient to interfere with the function of the apparatus. Enclosures to IP 6X are termed dust tight and should be able to exclude all observable dust particles.

The temperature specified for the free surfaces in these standards allow for dust deposits up to 5 mm thick.

An alternative division of risk used by German industry closely follows the IEC and CENELEC hazardous area classification (HAC) used for explosive gas atmospheres. Two zones are specified:

Zone 11 (Corresponding to HAC Zone 1)
Where an explosive dust atmosphere will occasionally occur for brief periods due to deposited dust being stirred up

Zone 10 (Corresponding to HAC Zone 0)
Where explosive dust atmospheres are present for long periods or frequently.

The required ingress protection here is IP 54 for Zone 11 and IP 65 for Zone 10.

The temperatures specified for free surfaces allow for dust deposits up to 50 mm thick, the maximum safe values being given by curves as shown in Figure 8.3.

Under the American national electrical code (NEC), the dust zones are of a similar nature to the German definitions and are termed Class II, Division 2 and Class II, Division 1, the respective enclosures for these two levels of risk being dust protected and dust tight, as specified in the Underwriters Laboratories standard 674. For the maximum surface temperature, the NEC assumes a depth of deposited dust of up to 12·5 mm.

Because of the difference between established American practice and European views in this field, the International Electrotechnical Commission has proposed

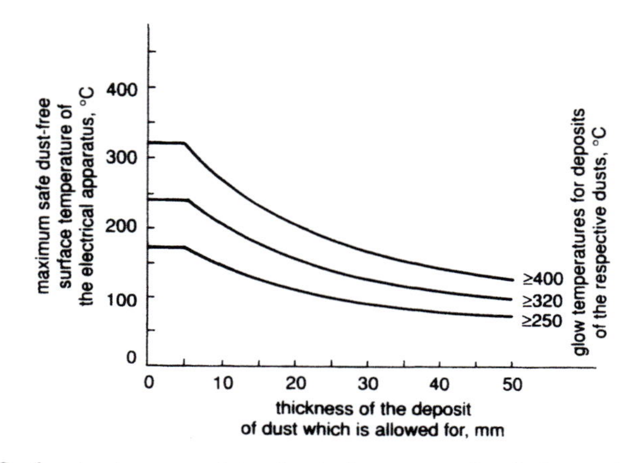

Figure 8.3 Surface temperature of the electrical apparatus in relation to the thickness of the dust deposit and the glow temperature of the dust

Limiting values as specified by DIN VDE 0170/0171 Part 13

what is in effect a hybrid standard incorporating protection form E (European) and protection form A (American).

With protection form E, the maximum surface temperature of the enclosure is determined without dust deposits and must be at least 75 K cooler than the minimum ignition temperature of the dust concerned when in the form of a 5 mm thick layer. The free surface temperature must also be not more than two-thirds the cloud ignition temperature (this presumably applies to the 0–100°C scale). These requirements have been widely adopted in national European standards.

With protection form A, the surface temperature is measured with not less than a 12·5 mm thick layer of dust on the enclosure and the measured value must be 25 K below the ignition temperature for the 12·5 mm deposit and again not more than two-thirds the cloud ignition temperature.

All the above practices can be regarded as providing a factor of safety, but the UK standards are rather limited by ignoring requirements for deposits of more than 5 mm. This is somewhat unrealistic in many applications such as cyclone bagging and the carding of raw textiles.

8.5 Causes of dust fires and explosions

According to the records of HM Inspectorate of factories, relatively few fires and explosions involving flammable dusts, fibres etc. have been caused by electrical equipment in normal operation. Over a twenty year period, out of nearly 1000 dust fire and explosion incidents, only 23 could be definitely attributed to electrical installations. Six of these were reported as explosions and 17 as fires. Three of the explosions were associated with incandescent light bulbs breaking within a dust cloud. On this basis one can say that light bulbs are a significant cause of dust fires and explosions.

8.6 Bibliography

PALMER, K.N., *Dust explosions and fires*, Chapman & Hall, 1973.
Dust explosions in factories, Health and Safety Executive, Publ. HMSO, 1975.

Chapter 9

Design, workmanship and maintenance

The failure or breakdown of electrical equipment can sometimes be blamed on faulty design or poor workmanship. Often, but not necessarily more often, it is the result of inadequate maintenance. We will begin by considering some effects of faulty design.

The commonly held view is that the main hazard with electrical installations is the risk of electrocution. This is certainly not the case. The main hazard is fire. The number of people electrocuted each year in the UK is about the same number as are killed by lightning. That is not to say that electric shocks are uncommon. Probably about a million a year experience some form of electric shock, yet less than 50 are killed in this way. Electrocution requires a potential through the torso of not less than 50 volts whereas fires can be caused by almost any voltage. Most automobile fires are due to 12 V electrical faults accompanied by leaking fuel. They occur too frequently to have much news value although the results can range from expensive to lethal, or to ghastly second and third degree burns.

Where there is a design weakness on mass-produced products, even very slight hazards become important. New automobile models occasionally have to be recalled in their thousands for minor modifications due to extremely remote hazards, not perceived until a few accidents are reported. Wiring which becomes overloaded due to component failure or becomes short-circuited due to chafing will in most cases cause a fuse to blow with no consequent danger of fire. Statistically, in spite of fuse protection, about 0·1% of these faults will lead to a conflagration and the number of fires will be proportional to the number of such cars on the road.

Other instances where the smallest chance of failure is intolerable are those involving possible nightmare scenarios. In these cases the specification may require safety to be preserved with several simultaneous independent faults. Where one fault can lead to another, the combined effect must be regarded as one independent fault.

It is normal practice to design mains-fed installations and equipment to be safe, as regards electric shock, with up to one fault. The risk of fire is treated differently in that electrical equipment is normally designed so that a single fault may be capable of causing a fire. For example, hidden wiring to a switch, when disturbed or pierced by a nail, can ignite adjacent woodwork and other flammable materials without drawing enough current to operate either overload or earth-fault protection. The fire which destroyed a large section of Windsor Castle in 1992 was also ignited by one fault from an electrical device — a lamp placed and left too near a curtain. It can be argued that, where wiring or apparatus is able to cause a fire while in service as a result of a single mistake, there is a fundamental weakness in the design.

9.1 Designing for safety

With regard to electrically driven machines, the prime hazard is the possibility of damage or injury caused by the driven machine. Safety requirements are therefore mostly concerned with being able to stop the moving parts as quickly as possible in an emergency. Simply switching off the supply to the motor is sometimes too slow to avoid an accident and braking, or even reversing, may be necessary. Stopping devices for the driving motor must be readily accessible, easy to operate and, if the operating positions require it, more than one stopping device may be necessary — each of which has to be manually reset after any one has been operated. Motor stop buttons should be of the mushroom-headed type and be coloured red with a yellow background. Other methods of arresting movement can be provided by a tensioned trip wire attached at one end to a pull switch, as used for conveyors. For some larger machines, trip barriers or pressure mats which will switch off the power if one tries to enter a danger zone will be suitable.

The guarding of machines applies mainly to manufacturing processes. Factory employees on piecework or production bonus rates are particularly vulnerable to injury and the guarding of machines to prevent these people from harming themselves needs special care. Their work positions must be arranged so that, as far as possible, the protection provided will (a) not fail in a dangerous mode, i.e. will fail safe and (b) cannot be defeated by the operator it is intended to safeguard. Among the most dangerous processes are hand-fed machines such as power presses, calenders and woodcutting circular saws. All these machines need protection and the guards must be rigidly secured and robustly constructed to ensure they will not yield if a person stumbles or falls against them. The standard type of guard over a circular saw, which hinges upwards at the front is a poor design in this respect. The guard can be forced upwards by the hand pushing the workpiece into the blade — with dire and messy results.

Guards for machines such as stamping presses, press-brakes and hand-fed calenders need to be interlocked with the drive mechanism. The interlock may be operated via the power supply to the motor, or through some mechanical disconnection device such as a gear or clutch engagement. If the closure of a barrier is arranged so that it forces a starting switch to the 'on' position, the system does not fail safe since the switch contacts could remain closed — due to welded contacts or a weak spring, for example — when the barrier is removed. Also the safety interlock can be readily defeated by placing a chock, instead of the guard, against the switch plunger. To provide a fail safe or positive mode of operation, the starting switch should be normally closed, i.e. should close by its own spring return, the barrier being designed so that when it is in place, the switch is free to close. The weakness here is that, if the guard is removed altogether or becomes distorted, the starting switch may be able to reach the closed position with the machine unguarded. This risk can be reduced by having a positive mode and a negative mode switch in series.

An essential element in machine guarding is to ensure that the movements have stopped before the operator can get past the barrier. Various ways of ensuring a sufficient delay can be applied. One of these depends upon the exchange of a key in a key-trapping box. For low inertia machines, a braking system is often sufficient. Power presses are frequently designed so that a knee-joint of the stroke mechanism collapses enabling the dangerous motion to be immediately arrested without having to stop the flywheel.

The use of limit switches to prevent excess travel of a machine also requires some care in the general design. The switch should not be in the line of travel where it can be crushed if its instruction fails to be obeyed. Microswitches are not entirely suitable as limit switches but can be used to detect the position of a machine part where personnel safety is not involved. Figure 9.1 shows the correct and incorrect mounting positions for a microswitch fulfilling the function of a proximity detector.

Instead of mechanical barriers to guard machines, so-called intangible barriers can be adopted. These are generally electrosensing systems which detect any dangerous intrusion by sensing a change in the control circuit. Depending on the application, detection systems can be based on photo-electricity, capacitance, inductance, microwaves, radar, infra-red radiation and ultrasonic acoustics—virtually the same choice as for intruder alarms. Because the barrier cannot be seen, heard or felt it must be even more reliable than a physical barrier. Special high integrity circuits which can only fail in a safe mode are required, but where a certain time lag is necessary to ensure safety, the use of electronic surveillance may not be applicable.

Modern electrical practice requires electronic control and signalling circuits for 3-phase motors to be supplied from an isolating transformer, usually energised across two phases of the supply network where no neutral is present. It is important that earth faults on motor control circuits are not able to cause accidental starting or to prevent the machine from being stopped. The preferred circuit is to have one side of the control power source earthed, with the control switch on the live side of the contactor coil. As can be seen from Figure 9.2, no earth fault on the control circuit can then cause spurious operation.

In the design of electrical control cabinets, it is often convenient and sensible

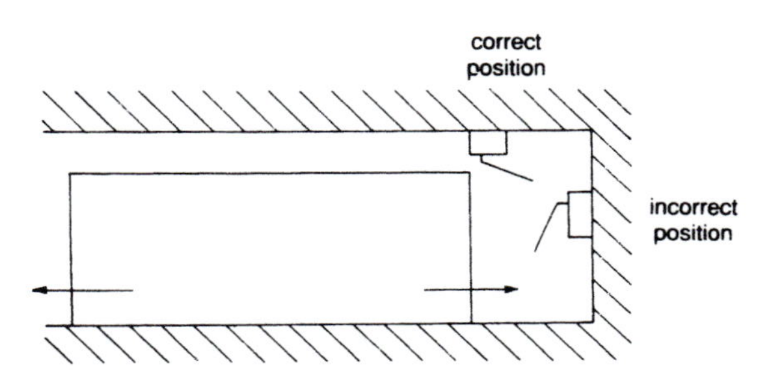

Figure 9.1 Correct and incorrect positions for a microswitch or proximity switch to detect the position of a moving part

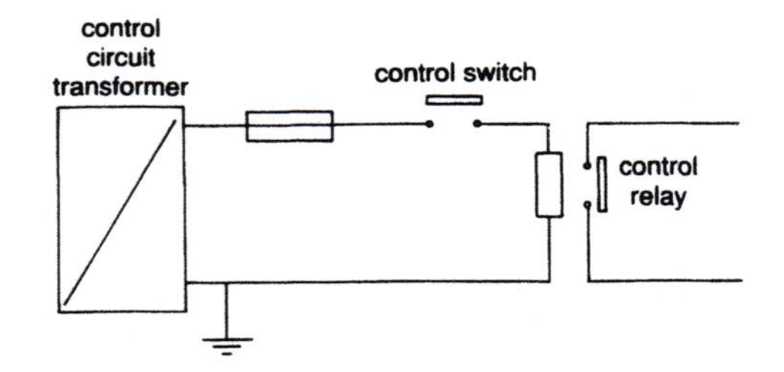

Figure 9.2 Preferred control switch circuit for electric power drives

No earth fault on the control circuit can cause the relay to operate

to mount some of the interfacing components on the door or hinged cover of the enclosure. Items of this sort will comprise indicating lamps, instruments, selection switches, trip reset and alarm cancellation buttons. If the replacement of lamps and fuse links requires access to the rear of the door, it may be inconvenient to interlock the opening of the door with the main power supply. In this case, the equipment within the enclosure and on the rear of the door will need to have protective screening over bare live parts and terminals to prevent accidental direct contact when the door is open.

Door-mounted equipment will also require a suitable flexible cable and protective conductor across the hinge. This cable should be secured at each end independently from the termination of its conductors and should be arranged so that frequent opening of the door presents no risk of damage or fraying. Bolted covers can normally be expected to be electrically connected to the enclosure by virtue of the fasteners, but hinged covers need to have a definite earth bond to the case if they carry energised items of equipment. For hinged covers with no electrical equipment fixed to the cover, brass or other non-corroding metal hinges can usually be regarded as a reasonable connection.

Withdrawable racks for electrical apparatus are commonly considered to be adequately earthed through their sliding contacts with the housing cabinet. Where the racks are connected by multi-pin plugs and sockets, there is no reason why the connector should not include a protective conductor for the chassis of the rack. As with all plugs and sockets, the PE circuit should be established before the live circuits and should be interrupted last on withdrawal of the connecting device.

On the question of potential equalisation conductors, there is always some doubt as to whether isolated items like metal rating plates pop-riveted to plastic enclosures need to have an earthing connection. An accepted rule is that no PE is required for parts which

(*a*) Cannot be grasped by the hand, or
(*b*) Have a surface area of less than about 50 mm × 50 mm, or
(*c*) Are located where they can never come into contact with live parts.

These guide-lines are all rather vague and general practice is to ensure that metal screw heads visible on the outside of plastic covers—on wall sockets and switch boxes for example—should be connected to a PE conductor within the box, even though the heads of the fasteners are recessed. The need to bond metal rating plates can be avoided by fixing them with an adhesive in place of metal fasteners.

Where small exposed metal parts are too small to be grasped, there is no question of not being able to let go if they happen to be touched while live. A shock from inadvertent contact will produce a spasm which immediately breaks the contact. Danger of electrocution from rating plates or from screw heads recessed into a plastic cover is therefore negligible except in particularly hazardous areas, where it is preferable in any case to avoid the installation of equipment having any exposed conducting parts. Light switch boxes without exposed screw heads and which are effectively Class II items should be selected for humid environments since even PE-connected metal depends on the existence of many invisible connections for its safety.

The reference to rating plates reminds us that durable and legible marking and labelling is an essential part of safe design. Motor rating plates are often fixed on the side of the machine which is close to a bulkhead or the back of a cabinet. The essential data are then virtually impossible to read, even with the aid of a torch and a mirror. In these instances duplicate plates should be fitted in a legible position. Cabinets and wall boxes containing electrical apparatus or just terminals should always bear an identification label indicating their purpose and voltages. All cable core and wiring terminations should be clearly marked in a way that can be read without having to disturb the connections.

9.2 Workmanship and maintenance

We are referring under this heading to amateur workmanship. Electrical firms and electricians approved and registered by the National Inspection Council for Electrical Installation Contracting (NICEIC) can be relied on to carry out work of adequate quality. But in spite of the general influence of poor workmanship on safety and the need to understand wiring rules, there is a tendency under British jurisdiction to allow householders of any age or ability to undertake their own wiring additions and repairs, whether it be fitting plugs or running a power supply to the greenhouse. In mainland Europe a high proportion of people build their own property, but this is usually undertaken by a consortium of professional skills on a neighbourly basis. There is no encouragement for amateurs to try their hand at domestic wiring and all electrical appliances designed to be energised from a power socket must be sold with moulded-on plugs. This rule will soon be universal as the danger associated with the incorrect connection of a plug top can be far more serious than just omitting the protective conductor. If the line and earth wires are reversed in the plug top, the casing of the appliance will stay at line voltage to earth whenever an attempt is made to use the equipment.

Bayonet and socket outlets which have been removed without disconnecting or insulating the bare wires supplying them are a frequent source of danger. There are two instances which come to mind: bare lighting leads hanging down below head height in the otherwise unlit bedroom of a lodging house, causing a new occupant a painful shock in the darkness; live wires left behind wooden panelling after a dockyard refit of tanker accommodation, causing a smouldering fire at sea.

Poor workmanship is always most prevalent where it is unlikely to be seen and the protective conductors are the most likely to be subjected to short-cuts.

Maintenance is important but can be misapplied. If preventive maintenance is undertaken only during a planned shut-down and consists merely of the invasion of electrical enclosures with a compressed air hose or the nozzle of an industrial vacuum cleaner, the mean time between faults is likely to decrease.

Moving parts need to be adjusted and, in principle, must eventually wear out. They do need periodic inspection. Equipment with no moving parts, such as terminal boxes and solid-state circuits are unlikely to deteriorate if used within their designed rating. Regular maintenance or cleaning here may, therefore, be less than cost effective.

This does not always apply. Mentioned below are notable exceptions where regular inspection and appropriate servicing of non-moving equipment is advisable.

In screw and tunnel terminals carrying a cyclic load, as when a high current device such as an immersion heater is controlled by a thermostatic switch, the repeated on/off cycle will cause copper conductors to expand and contract, leaving the connection slightly loose. Contact resistance then increases. This further promotes the ohmic heating effect and a rapid rise in temperature can occur, causing the wiring insulation near the terminals to be destroyed. This runaway condition is prevalent with aluminium conductors if the initial temperature rise goes above a certain level.

With sensitive electronic devices and circuits, damage can be caused by electromagnetic interference, possibly radiated from voltage spikes on the mains system near the apparatus. The screening and earthing need to be kept in good order.

Although batteries contain no moving parts, larger installations with wet cells certainly need to be regularly inspected and maintained in accordance with the manufacturer's instructions.

Where there are particularly aggressive or demanding environmental conditions, it is wise to ensure from time to time that the interior of apparatus is in a healthy state. The original specification of the enclosure should be in accordance with the circumstances of use. However, the working conditions may change. Furthermore, the actual ambient conditions may not match expectations.

Figure 9.3 is a view of an electric heater and control box for frying chips. Due to lack of hygiene and maintenance, the underside of the box was coated with burned and congealed fat and much of the same is in the interior of the control unit. The poor state of the equipment was revealed only after a member of the kitchen staff was injured, having received a very severe shock from the unearthed metal of the casing.

The need for inspection and maintenance of oil-filled transformers depends to some extent on their size and loading. Unless they are subjected to severe short-circuits or lightning strikes, pole-mounted units and distribution transformers in general can normally operate unattended for periods of a year or more. A regular examination of the state of the oil is still the best indication of the condition of the windings.

As the windings age they can release products which reduce the quality of the oil. In an old transformer, replacement oil may therefore not last as long as the original charge.

Some high voltage switchrooms in isolated buildings on the perimeter of industrial sites are never visited from one year to the next. Internally they may be warm and dry so that no maintenance or even periodic inspection is considered necessary. On entering one of these buildings, the visitor may find the switchboard instruments and relay glasses totally obscured by cobwebs and a smell of ammonia

Figure 9.3 Electric heater from a chip fryer

Inspection showed that a new cable with CPC had been fitted, but the interior of the control box remained coated with rancid fat

will indicate the presence of mice. If a window has been broken and left in that state for a year or more, the conduit and lighting switches will be rusty. There may not be spiders but the tops of the equipment will typically be covered with layers of bird lime. Unattended switchrooms should be inspected at least twice a year. Light fittings are among the first items to reveal signs of neglect. An example is shown in Figure 9.4, admittedly a special case.

Most maintenance is carried out after a functional failure of some sort. This is not really good enough as overheated or faulty insulation may only announce itself by an electric shock or the presence of smoke and fire. There is the counter-argument which says 'If it hasn't gone wrong don't repair it', which is supported by the statistical knowledge that the majority of functional failures can be traced to the most recent occasion when the equipment was serviced. A good policy for the pre-emption of equipment breakdown is periodic inspection with a clipboard and a torch. On average, 98% of items which need attention can be found by critical observations – in the visual, not the verbal sense. The eyes of an experienced technician can collect a great deal of information regarding the health of an electrical installation, without needing to switch off or to remove anything. The only other essential requirement is a critical state of mind. It is also relevant to remember when making any examination or tour of inspection that it is more difficult to identify what is missing than to question what is present.

Regular visual plant inspections can monitor at least some of the following:

- Broken or missing earthing and bonding cables
- Lights not working or flickering

Figure 9.4 Lighting installation in an illicit factory

- Lighting covers cracked, broken, missing, dusty, oily or burned
- Covers missing
- Bolts or nuts missing
- Guards broken, loose, displaced or missing
- Oil or water leaking
- Instrument glasses broken or cracked
- Instruments unreliable, reading incorrectly, or stuck
- Conduit not connected into equipment
- Exposed wiring, e.g. split cable sheaths
- Boots on cable glands split or disturbed
- Cable runs displaced
- Signs of overheating, rust etc.
- Need for cleaning, greasing, or painting
- Damaged equipment
- Labels illegible, incorrect, or missing
- Poor housekeeping, e.g. untidiness and combustible litter

- Unsafe access for operation and maintenance
- Unexplained or unusual noise/smell

A hands-on inspection can also check on cables

- Are they tight in their glands?
- Are straps and clamps secure?
- Are they running hotter than usual?

and machines

- Is there any undue vibration or noise?
- Has it got worse?
- Do the bearings feel too hot?
- Are the oiling rings rotating?
- Do the machine frames feel too hot?
- Are the fan blades reasonably clean?
- Are the cooling air ducts choked?
- Are drains and breathers working correctly?

A planned shut-down inspection can check

- Condition of electrical contacts
- Tightness of terminals and connections
- Presence of dust, water, oil etc. in housings and equipment
- Lamp ratings in lighting fittings
- Presence or evidence of condensation in enclosures or machines.

It is worth noting that none of the above inspection checks requires any instruments whatsoever.

Chapter 10
Stored energy

10.1 Batteries

Stored energy is not hazardous until it is released or transformed. This applies in a mechanical context, as when falling out of a window, in an electrical context, as when being struck by lightning and in a chemical context, as when stepping on a land-mine. In each of these examples the potential energy is discharged very rapidly. Batteries are devices for storing chemical energy which can be released at a controlled rate in the form of an electric current. The maximum rate of release—when the terminals are short-circuited—will be limited by the internal resistance of the battery.

The internal resistance of a battery is thus one of its three most important characteristics, the other two being its maximum stored energy/unit volume and maximum stored energy/unit weight. These parameters apply to both primary (non-rechargeable or throw-away) devices and secondary (rechargeable) devices.

Other factors have to be taken into account in the design and selection of energy storage cells, depending on their application, but the following are of increasing environmental and commercial importance:

- Disposal costs
- First cost
- Explosion and chemical hazards
- Toxic effects
- Maintenance requirements
- Charge retention
- Charge/discharge efficiency.

The first of these applies mainly to primary cells, many thousands of millions of which are disposed of every year, with production increasing exponentially.

Count Alessandro Volta, who invented the first machine to generate static electricity by induction, was also the first to describe an electrochemical battery—in a letter to the Royal Society in 1800. Volta's pile consisted of a stack of zinc and copper discs separated by felt moistened with dilute sulphuric acid. The concept of a pile of dissimilar plates has been perpetuated in the Gallic term, une pile, for an electrical battery. Today, some hundreds of different material combinations are being actively explored and developed. Among the most commercially important types are

- Lead/acid batteries
- Nickel/cadmium and nickel/iron alkali batteries
- Lithium anode batteries
- Leclanché batteries.

10.1.1 The lead/acid battery

This type was first made in 1859, by the French physicist Gaston Planté (1834–1889). Although the earliest rechargeable electric storage system, it is still the basis of the most widely applied secondary battery, as used universally for road vehicle starting, lighting and ignition (SLI) duties. The extensive adoption of this type of battery for automobiles absorbs about a quarter of the world's output of lead.

Each cell of a lead/acid battery consists of a positive electrode of lead dioxide PbO_2 paste and a negative electrode having an extended surface of spongy lead, the two electrodes being permanently immersed in dilute sulphuric acid H_2SO_4. During discharge, the electrochemical reaction, which is reversible, can be written as

$$Pb + PbO_2 + 2H_2SO_4 = 2PbSO_4 + 2H_2O$$

It will be seen that the constituents are converted to lead sulphate and water so that, as the discharge proceeds, the acid becomes more and more dilute. Both the density of the electrolyte and the terminal voltage on open circuit decrease linearly as current is drawn from the battery, giving a useful means of measuring the state of discharge of an individual cell and the overall residual capacity of the battery, respectively. At the same time the internal resistance rises due to deposition on the electrodes of lead sulphate, which is insoluble and a poor conductor.

During recharging, the chemical process is reversed and once the battery has become fully charged the water content of the acid solution starts to be electrolysed and dissociates into gaseous oxygen at the positive electrode and hydrogen at the negative. Continued over-charging therefore simply reduces the water content of the electrolyte so that topping-up with distilled water becomes necessary.

Batteries used for starting petrol and compression-ignition engines are required to supply a high power for brief periods – about 4 HP (or 3 kW) for vehicles with diesel-cycle engines and somewhat less for petrol car engines. For petrol engine starting, the voltage must in addition not fall unduly since the spark-plugs also have to function during the cranking period. The low internal resistance which can be obtained with the lead-acid battery is thus one of its essential advantages for SLI duties. Its chief disadvantages are

- Its high weight for a given capacity, due to the lead content
- Its inability to retain charge for more than a few months
- Its irreversible loss of capacity if left in a discharged state for an extended period due to the lead sulphate on both positive and negative electrodes.

With suitable charging and maintenance, a lead/acid battery can be recycled several hundred times and will last for several years as a back-up power source on float charge. The overall life is usually determined by the loss of the lead oxide paste from the positive plates. This falls to the bottom of the case and may eventually cause the electrodes in one or other of the cells to become short-circuited. Accordingly, the life of an SLI battery is likely to be reduced by excessive vibration.

Various additional metals are used to improve the mechanical properties of the electrodes, a lead-antimony alloy having in the past been commonly incorporated in the design. For road vehicle SLI service, maintenance-free batteries are now in general use. These effectively eliminate loss of water from the electrolyte by enabling the dissociated oxygen and hydrogen to recombine within the casing, usually with the aid of a catalyst. The complete battery can then be sealed to prevent evaporation. Even so, pressure relief safety vents are always necessary to ensure against an explosion of the case in the event of severe over-charging.

The single car battery does not normally present much risk, although they can cause injury if misused or carelessly handled. With larger installations, such as for electric vehicles, electric traction and standby power supplies, there are four main hazards associated with lead/acid batteries.

10.1.1.1 The electrolyte

Sulphuric acid is a corrosive, poisonous and dangerous liquid which in a concentrated form instantly chars wood. Even the diluted acid used in batteries will rapidly destroy clothing. Goggles and plastic apron must always be worn when dealing with lead/acid battery installations. Any acid on the skin must be washed off immediately. If electrolyte gets splashed or sprayed into the eyes it will probably result in injury to the sight as well as being excruciatingly painful. When examining or working on a battery installation consisting of open cells, a suitable quantity of alkali such as soda ash or washing soda should always be available to neutralize and wash off any splashed or spilt acid. When filling cells which have been delivered without electrolyte it is very advisable to wear full acid-proof personal protective equipment (PPE), i.e. face and eye protection, rubber boots, gloves and apron.

If it is necessary to dilute concentrated acid, at least two operators must be present and a plentiful supply of clean water must be immediately available. Having said that, it must be remembered that water should never be added to a vessel containing concentrated sulphuric acid. The large amount of heat generated can turn the first drops of water to steam causing violent ejection of the acid from the vessel. Dilution should always be carried out by slowly pouring the acid into the water while carefully stirring the mixture.

10.1.1.2 Gassing

When batteries of this type are given a periodic conditioning charge, hydrogen and oxygen are generated in the most explosive (stoichiometric) proportions. The hydrogen content will cause the mixture to be lighter than air so that it will tend to rise. Battery room ventilation should accordingly be at high level. Where several banks or strings of cells are connected in parallel, the failure of one cell will reduce the voltage of the bank in series which will then be overcharged by the parallel groups. An automatic charging system will try to compensate for the overall loss of voltage leading to continuous overcharge of the group containing the faulty cell. As a general rule, lead/acid batteries will start to generate an explosive mixture of hydrogen and oxygen when the voltage on charge rises above 2·30 V/cell. For this reason, installations consisting of parallel banks of batteries, such as are used for uninterruptible power supplies (UPS) for computers, process controllers, file servers etc. should be installed in separate rooms having a plenum ventilation of not less than 6 changes of air per hour.

Due to the possibility of an explosive atmosphere, battery rooms and cabinets should always be regarded as non-smoking areas. When enclosed batteries are tested with a load device having test spikes, there will usually be a spark as contact is made and broken. If the battery has been gassing and hydrogen and oxygen are escaping from the cells, or from a single faulty cell, the spark in the vicinity of the vent could ignite the gas and cause the battery case to explode. Before testing with load spikes a battery must therefore be disconnected and freely ventilated.

We have mentioned that the internal resistance of a battery rises as it discharges due to the deposition of lead sulphate on the electrodes, lead sulphate being insoluble and a poor conductor of electricity. If a battery is left in a discharged condition, its internal resistance will continue to rise due to further formation of the sulphate and eventually it will not be possible to restore the battery to a charged condition with any useful ampere-hour capacity.

Attempts to recharge a flat battery by a rapid or boost charge are dangerous. Forcing a high charging current into a fully discharged lead/acid accumulator will produce a voltage per cell above the 2·30 volt gassing level. Badly sulphated cells will not recharge at all, but if current is passed through them, the water content of the electrolyte will be dissociated into gaseous hydrogen and oxygen. The effect will be to dry out the battery without restoring it to the charged state. The liquid level will fall leaving partly empty cells containing an extremely explosive gas mixture. The temperature may also become excessive, causing the remaining acid to boil and possibly buckling the plates. If they touch, cathode to anode, inside the battery, or the liquid level falls below the bottom of the plates—which do not normally extend to the full depth of the casing—there will be a spark igniting the hydrogen and oxygen and producing a violent explosion. This will certainly burst the battery case and could severely injure anyone in the vicinity by expelled acid and flying debris. Cells having a high internal resistance should be recharged at low currents until the resistance comes down.

Car and lorry batteries placed on charge may have one dud cell with a high resistance and this will gas even when the other cells are charging normally. Spring-clip connections from the charger to the battery terminals should therefore never be removed without first switching off the supply to the charger.

10.1.1.3 Toxic fumes

These may be given off during charging due to the presence of antimony in the lead alloy plates which produces the poisonous gas antimony trihydride SbH_3. For this reason modern designs are tending to use alternative alloys. However, for the reasons given above, there is still a need for batteries and battery rooms to be well ventilated.

10.1.1.4 Short-circuits

The SLI battery is designed specifically to deliver 200 A or more with only about 25% volt drop. Its short-circuit current is therefore going to be a value approaching 1000 A. We have to bear in mind also that, if the battery terminals are shorted by a wire or metal tool for instance, there will be no fuse or circuit-breaker to interrupt the current. Consequently, only insulated tools should be used when working on or over batteries. It is also very advisable to remove metal watch straps, necklaces, rings and any other metal accessories since a short-circuit current through metal in contact with the skin can produce a serious and painful burn—possibly with the metal fused to the skin. Even with enclosed battery cases where there is virtually no risk of splashing from electrolyte, goggles should be worn to protect the eyes from flash and metal sputter which can occur as a result of an accidental short-circuit across the battery terminals.

Lead/acid batteries can be designed to have a capacity of about 40 Wh/kg and 75 Wh/litre size although their performance is significantly reduced at low temperatures.

10.1.2 Nickel/cadmium and nickel/iron batteries

The nickel/cadmium battery is based on an 1899 patent by Waldemar Jungner of Sweden. It consists of a negative electrode of metallic cadmium whose surface is extended in some way such as by means of a porous sintered construction, and a positive electrode of pelletised or sintered nickel hydroxide $Ni(OH)_2$ powder. The electrolyte is an aqueous solution of potassium hydroxide KOH. On discharge, the nickel hydroxide converts to a hydrated oxide NiO(OH) with the liberation of some oxygen which subsequently becomes reabsorbed. The reversible electrochemical process on discharge can be shown as

$$Cd + 2NiO(OH) + 4H_2O = Cd(OH)_2 + 2Ni(OH)_2 \cdot H_2O$$

The following list outlines the advantages of the nickel/cadmium cell.

- With float charging there is no loss of electrolyte. Some water loss will occur during the periodic conditioning charge but, by having a large reserve of electrolyte, topping-up is only necessary after long periods of service.
- For portable and miniature (button) batteries, sealed cells can be used and operated in any position.
- The cells, which are usually individually encased either in metal or transparent plastic, can be operated at high charge and discharge rates.
- Their open-circuit voltage of approximately 1·3 V remains substantially constant during the charge/discharge cycle.
- The potassium hydroxide plays no part in the reaction and accordingly the electrolyte concentration remains the same throughout the cycle. This means that, in the discharged condition, the battery is less subject to freezing than a lead/acid battery.
- It also has a better low temperature performance.
- At normal temperatures, the charge retention is better, being a matter of years rather than months.
- No permanent loss of capacity is incurred if the battery is left for long periods in a discharged state.

- The chemical hazards from the possible release of toxic gases are less, although the electrical hazards will be much the same as for lead/acid installations.

Because of their long life, robust construction, reliability and insignificant maintenance requirements, the nickel/cadmium battery is widely used for remote or unattended installations such as for the automatic starting of large stationary engines driving pumps and standby generators. Their ability to function without attention as small sealed units has given them a market in cordless power tools and as pcb-mounted electronic back-up power sources.

The chief disadvantage as regards the larger installations lies in their first cost, which is normally several times higher than an equivalent lead/acid battery. Part of this cost is associated with the health hazard incurred in the handling and processing of cadmium.

Because the specific gravity of the electrolyte and the open-circuit voltage remain virtually constant, there is no simple way of measuring the residual capacity at an intermediate stage of discharge, as there is with lead/acid cells. The onset of the fully discharged state will obviously be indicated by a rapid loss of voltage and the completion of the charging period is denoted by a rise in voltage above the normal 1·3 V/cell.

While the electrochemical process can be shown as a simple conversion of the metallic cadmium to cadmium hydroxide, there is in fact a complex intermediate stage which requires up to 50% more energy for the charge than is delivered during the discharge. The cycling efficiency is therefore less than can be obtained with a lead/acid battery. But, for the applications indicated above, the lower efficiency is not of great importance in relation to the other advantages.

As mentioned, it is occasionally necessary to restore the level of electrolyte by the addition of water. This will be a maintenance requirement in the case of the larger batteries which generally need to be vented as a precaution against over-pressure. Smaller cells will normally be fitted with a pressure relief device. Apart from this, the only other servicing that may be needed is to replace the electrolyte every thirty years or so due to the formation of carbonate. This is caused by the potassium hydroxide combining with atmospheric carbon dioxide CO_2 to form potassium carbonate.

The Wh/kg available from a nickel/cadmium battery is comparable with that from a lead/acid battery.

Nickel/iron batteries are very similar to the nickel/cadmium designs and were patented by Edison in the USA at about the same time as Jungner's nickel/cadmium patent. The electrolyte and positive electrode are the same but the negative electrode is of iron in finely powdered form. In this case the reversible electrochemical reaction can be represented on discharge by

$$Fe + 2NiO(OH) + 4H_2O = Fe(OH)_2 + 2Ni(OH)_2 \cdot H_2O$$

These cells are also of rugged construction and will provide a similar amount of energy per kilogram. Their main disadvantages are: the reduced cycling efficiency and a poor charge retention, their rate of self-discharge being somewhat more rapid than for lead/acid cells.

Until fairly recently nickel/iron batteries were not widely manufactured although it has long been known that they have considerable commercial advantages for electric vehicles. At the beginning of the 20th century many alternative forms of road transport were being actively promoted under the general title of automobilism. Energy sources included gas, steam, oil and electricity. Even in 1898, the French Automobile Club was carrying out battery trials at a Concours d'Accumulateurs. In the summer of 1903, an Edison nickel/iron battery of 38 cells was used for tests on a Studebaker car and was the subject of a paper presented to the Institution of Electrical Engineers in November of that year by W. Hibbert. The battery output was 11·8 Wh/lb (26 Wh/kg) and the cycling efficiency about 60%. The overall battery capacity of some 8 kWh remained the same after trial runs in France and England totalling over 900 miles and the cells were subsequently installed in a London cab. An extract from the trial records, reproduced in Figure 10.1 shows that at this early date such batteries were already capable of practical use in electric vehicles.

In both the nickel/cadmium and the nickel/iron cells, the aqueous solution of potassium hydroxide KOH, or caustic potash is, as its name implies, highly corrosive and toxic. The same precautions are therefore essential when dealing with alkaline batteries as for lead/acid batteries, except that soda ash or washing soda will not be appropriate as neutralising agents. Goggles must of course be worn when handling potassium hydroxide and any skin burns should be washed with clean water and covered immediately with dry gauze. The potassium hydroxide can be supplied in solid form for dilution in pure (distilled or demineralised) water. As with sulphuric acid, water should never be added to the pure hydroxide on account of the heat which is generated.

With nickel/cadmium batteries, float charging allows the battery to be maintained at about 80% of its fully charged state and does not necessitate any regular conditioning charge unless the fully charged battery condition is required. If the battery is on a discharge-charge cycling duty, an input from the charger to the battery of about 160% of the battery capacity is needed to restore it to

The following is a diary of our runs:

[1903]

Aug. 29	Standing discharge: 159 Ampere-hours	
Aug. 30	Paris to Versailles and back, through the Park of St. Cloud. Good climb. Run about Paris...	39 miles
Aug. 31	Eighteen miles towards Rouen and back. About Paris...	48 miles
Sept. 1 Sept. 4	Journey to London. Car ran across Paris, then train to Havre; train also from Southampton to Waterloo. Motor overhauled at Niagara Garage, then finished the discharge by running round London. At Southampton the battery had to be partially discharged through wire...	29½ miles
Sept. 5	London to Northampton, stopping at Dunstable for a partial charge. Part of the discharge was taken next day round about Northampton...	77½ miles

and so on

Figure 10.1 Extract of a paper presented to the Institution of Electrical Engineers, by W. Hibbert on 26 November 1903, entitled 'The Edison accumulator for automobiles'

the fully charged condition. This gives rise to some gassing and loss of water from the electrolyte. For applications where the battery is regularly discharged and recharged, it will thus be necessary to have improved ventilation and a more frequent check on the electrolyte levels.

10.1.3 Lithium anode batteries

The metal lithium has a number of special properties. It is the lightest of the metals and, being about half the density of water, is lighter even than paraffin wax. It has the highest specific heat of any solid element and its chloride is one of the most hygroscopic substances known. Like sodium and potassium, lithium reacts instantly with water.

Its special characteristics as regards voltaic cells are twofold:

- It has a higher electrode potential than any other anode material. Its potential against hydrogen which is taken as the datum is more than minus 3 volts. If it could be used with a fluorine cathode, whose potential against hydrogen is plus 2·87 volts, a theoretical open-circuit voltage of over 6 volts per cell would be available.
- It has a higher theoretical energy storage capacity in electrochemical terms than any other material.

It has long been realised that if a cell could be constructed with a lithium anode it could have a number of superior properties, namely lightness, high potential/cell and high specific capacity. Accordingly, a great deal of development work has been invested since the 1960s in attempting to exploit these properties of lithium. The earlier difficulties lay primarily in the high chemical activity of lithium and the need to find suitable anhydrous electrolytes.

When a cell is supplying a load through a resistive conductor such as a filament lamp or starter motor, a current of electrons passes through the load from the cathode to the anode. Within the battery the accumulation of charge, due to the electrons entering the anode, is counteracted by positive ions passing from the cathode to the anode through the electrolyte. For the cell to operate as a source of energy, the electrolyte must in effect be capable of transmitting positive ions but behave as an insulator with respect to electrons. If the electrolyte were able to pass both ions and electrons with equal facility, the cell would effectively short-circuit itself. The electrolyte is thus a critically important component of a voltaic cell.

The cathode is equally important and vast numbers of cathode/electrolyte combinations have been tried with lithium anodes. Many of these systems are now fully developed and commercialised. Cathodes which have been used in conjunction with lithium anodes include

- Sulphur dioxide SO_2
- Thionyl chloride $SOCl_2$
- Vanadium pentoxide V_2O_5
- Silver chromate Ag_2CrO_4
- Iodine poly-2-vinylpyridine $nI_2 \cdot P2VP$
- Manganese dioxide MnO_2
- Molybdenum disulphide MoS_2

- Copper oxide CuO
- Iron sulphide FeS_2
- Titanium sulphide TiS_2

Two of the most interesting are Li/SO_2 with an energy density of 450 Wh/litre and a cell voltage of 2·9 and $Li/SOCl_2$ with an energy density of 660 Wh/litre and a voltage of 3·6/cell. These figures compare well with the long-established Leclanché battery which gives 150 Wh/litre and 1·5 V/cell.

Hazards associated with lithium cells are chiefly concerned with their high concentration of energy. The maximum power—that is to say, rate of energy rather than total energy—which can be delivered by a cell will be when the on-load terminal voltage is half the open-circuit voltage. In this condition the load resistance is equal to the internal resistance, with maximum power take-off corresponding to matched load impedance. It is normally the case that, the faster a battery is discharged, the lower the available capacity in ampere-hours. Hence, maximum power take-off is unlikely to be the optimum discharge condition unless the cells happen to have a very high internal impedance. With a matched load, the heat generated in the battery itself will be the same as the energy donated to the load. Consequently, with high performance cells and with dry cells in particular the losses within the cell during fast discharge conditions will readily produce undue heating. Many of the newer types of miniature batteries can reach a high temperature and high internal pressure if subjected to a short-circuit in a charged condition. Lithium cells need to be sealed to exclude moisture and so require a pressure relief device to avoid the possibility of the capsule exploding. Organic electrolytes enable lithium primary cells to operate over a very wide range of temperatures, with a shelf life of some 15 years. However, even with the incorporation of suitable conducting salts, some of these batteries have a high internal resistance and are mainly used for micro-amp duties such as for data storage back-up power and for cardiac pacemakers.

The lithium anode has a promising future for long-life, low drain primary batteries and various solutions to the problems of harnessing lithium in primary cells have now been found and successfully marketed. The emphasis must now be on comparable development of rechargeable lithium anode cell formulations. The primary cells often contain highly toxic substances and any devices marketed for the purpose of recharging non-rechargeable batteries should be treated with great caution as any attempt to recharge a primary cell, whether sealed or vented, can be very dangerous.

10.1.4 The Leclanché cell

Although not the first voltaic cell to be invented, the Leclanché cell is one of the oldest systems. It was devised by the French chemist Georges Leclanché (1839–1882). Despite many subsequent improvements, the essential component materials remain basically as he described them in 1866. The anode is a carbon rod, the cathode a zinc rod and the electrolyte is an aqueous solution of ammonium chloride NH_4Cl (sal ammoniac). Although numerous alternative types of primary storage battery have been developed since the nineteenth century, batteries depending on the Leclanché principle still represent far and away the largest share of primary cell production. Sales of these cells in the USA alone exceed $1000 million per year.

Simple voltaic cells can be made from any number of dissimilar metals and aqueous salts. Their general disadvantage is that they do not give a constant current. When a load is connected, the current starts to fall and reduces to a very low value after a few minutes. The cause is the production of hydrogen which forms a layer attached to the anode. This layer sets up a back emf within the electrolyte, the effect being known as polarisation. In the Leclanché cell manganese dioxide MnO_2 is packed round the carbon anode to act as a depolariser by oxidising the hydrogen layer. The manganese dioxide is a poor conductor and powdered carbon is therefore mixed with it. The depolarising action, however, is slow and a Leclanché battery is not suitable for supplying a large current for a long period without a rest. If the load is switched off for a while the voltage will slowly rise and the battery can in this way be applied to repeated intermittent duties, as for instance alarm bells and occasional signalling.

The early cells, many of which are still in service, were arranged with the liquid electrolyte in a square glass jar of standardised dimensions. In the so-called dry batteries, which can be used in any position, the NH_4Cl electrolyte is made into a paste with water, zinc chloride and gum. A simplified section of a single cell dry battery is shown in Figure 10.2.

The exact chemical processes that take place within a Leclanché cell are complex and are not fully understood even today. The final reactions can be simplified to

$$Zn + 2MnO_2 + 2NH_4Cl = 2MnO(OH) + Zn(NH_3)_2Cl_2$$

and $$Zn + 2MnO_2 = ZnO \cdot Mn_2O_3$$

and $$2MnO_2 + 2H_2O + H_2 = 2Mn(OH)_3$$

Leclanché cells have a limited shelf life; also their low temperature performance is very poor. Considerable improvements have been made by continual competitive development over the last century and during this period their specific capacity has been increased by at least an order of magnitude. Nevertheless, the

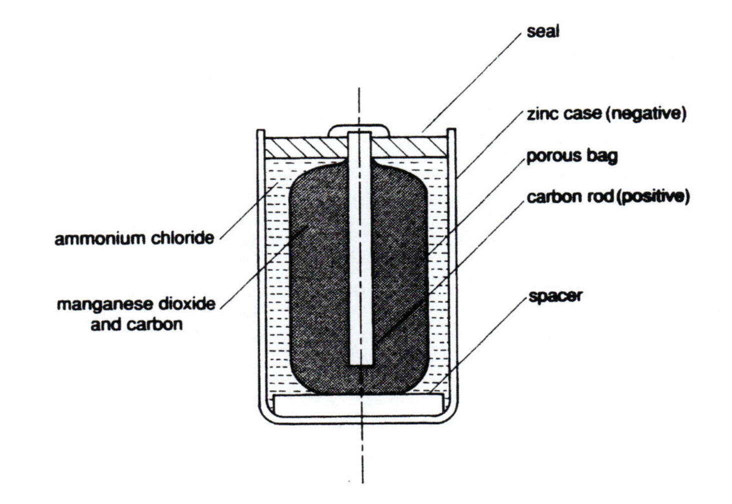

Figure 10.2 Elements of a Leclanché type single cell 'dry' battery

manufacture of the classical Leclanché dry battery is still an art rather than a science. The chief reasons for this are that both manganese dioxide and carbon exist in an indefinite variety of allotropic forms which have differing physical and chemical properties, depending on their provenance and subsequent treatment. The only proof of the effectiveness of a particular batch of manganese dioxide as battery material is by its behaviour in the cell.

Safety considerations with the standard dry battery are not as serious as with many of the more recently formulated types and are primarily a matter of the safe disposal of the used product. These can explode if over-heated and some have been known to burst while in use. The chief practical disadvantage of Leclanché cells is their habit of leaking after a heavy discharge or when left for a long time in a discharged state. The tendency of old cells to leak can be attributed to increasing acidity in the electrolyte, leading to corrosion of the zinc electrode with consequent evolution of excess hydrogen. The resulting pressure then forces the electrolyte through the seals of the casing. Hydrogen can also be generated if the battery remains connected to the load after the manganese dioxide has been used up. Typical examples are batteries which remain connected to electric clocks for years on end. Ammonium chloride is a slightly corrosive substance and the exuded electrolyte can spoil the appearance of polished furniture and indelibly stain many other domestic materials. Considerable effort has been given to the elimination of the leakage problem and modern dry batteries are now largely immune to this weakness.

10.2 Capacitors

These are also devices for storing electrical energy and generally consist of two metal surfaces almost in contact but insulated from each other by a dielectric material, i.e. a material which can withstand electrical stress without breakdown. In some ways a capacitor is similar to a secondary cell or accumulator in that it can store and release energy and can be charged and discharged. A secondary cell will store and release up to a certain amount of energy at a substantially constant voltage and has a significant internal resistance. The voltage across a capacitor, however, is directly proportional to the stored charge, which is normally in the order of ampere-hours $\times 10^{-12}$, and has a negligible internal resistance. Its discharge on short-circuit will accordingly be in the nature of a very short, high current pulse oscillating at the natural frequency of the short-circuit loop. Because the voltage is proportional to the stored charge, it follows that the stored energy (joule or watt-seconds) is proportional to the square of the voltage. The relationship is given by the simple equation

$$E = \tfrac{1}{2}CV^2$$

E being the energy in joule, C being the capacitance in farad and V being the initial potential in volts.

Capacitors, or condensers as they are also termed, are used for a number of important duties some of which are indicated in Table 10.1.

In any of its functions a capacitor has really only two parameters: its capacitance or energy stored per volt and its maximum working voltage. These two parameters are in opposition since capacitance decreases as the surfaces applying the energy-

Table 10.1 Some uses of capacitors

Purpose	Typical applications
To counterbalance the electromagnetic energy in an inductive circuit (i.e. to improve the power factor of the load, so reducing the maximum kVA demand and also the kVA consumption)	Incorporated in the control gear of fluorescent lighting fittings or installed in consumers' premises where there is a large inductive load (e.g. in factories and supermarkets)
To create a rotating field	In single phase motors
To form part of a resonant circuit	In tuned devices (e.g. radio, TV and amplifier systems)
To bypass the higher frequency components of a complex wave-form	In loudspeaker cross-overs and all types of filter circuit
To retain a potential difference or d.c. bias in a circuit	For solid state switching in computers, data handling and process control systems
To suppress electromagnetic interference	Petrol engine ignition systems and commutator motors
To produce a fatter spark	Ignition systems

storing stress to the dielectric are moved further apart, whereas the breakdown strength of the dielectric decreases as the surfaces are moved closer together.

Coming now to questions of safety, capacitors which are subjected to dangerous peak voltages, i.e. above 50 volts, should be permanently connected to a discharge resistor across their terminals, otherwise they can retain a shock potential after the power supply is switched off. It should also be remembered that, although there is what is sometimes called galvanic isolation between the terminals of a capacitor, this does not necessarily provide safe isolation for alternating currents.

The lowest safe value of insulating impedance as far as electric shock is concerned is a value which will restrict the current to about 2 mA. The impedance of a capacitor is inversely proportional to its capacitance and to the applied frequency, in accordance with the equation

$$Z = (2\pi f C)^{-1}$$

with impedance Z in ohms, frequency f in hertz and capacitance C in farads.

At a supply voltage of 230 V/50 Hz, a 1 μF capacitor would pass

$$230 \times 2\pi \times 50 \times 10^{-6} \text{ ampere} = 73 \text{ mA}$$

which could be lethal. The maximum safe value of capacitance in this instance would be 0·02 μF which would allow about 1·5 mA to flow. At 5 kHz, even 0·02 μF would be dangerous as the impedance would then be 1% of the value at 50 Hz. The term galvanic separation therefore has to be used with some caution when it is applied to capacitors in alternating current circuits or direct current circuits containing harmonics (ripple).

For single-phase power supply capacitors, such as in fluorescent lighting fittings and electric motors, the dielectric is normally a plastic film, typically polypropylene, polyester or polycarbonate. The conducting surfaces are usually of aluminium or zinc, vacuum deposited on the film. For larger units, aluminium foil may be used. The construction is in the form of a swiss roll, with one electrode overlapping the film at one end and the other electrode overlapping at the other end. The dielectric films are as thin as can be sustained in order to obtain the maximum reactance within the minimum volume. Local breakdown can occur due to imperfections in the dielectric, but these can be self-clearing by virtue of the extremely tenuous flashing of metal forming the conducting surfaces. To achieve this self-healing feature it has been common practice to impregnate the capacitors with an insulating oil.

10.2.1 The PCBs

Until the 1930s, insulating oils used in transformers, capacitors and switchgear were inflammable, refined mineral oils from selected crude petroleum. Mineral oil is a good insulator when it is clean and fresh, but degrades and forms an acidic sludge in use. Besides being inflammable itself, when the oil is subjected to arcing, it breaks down into the explosive gases, hydrogen and acetylene. Then along came the Askarels or PCBs (polychlorinated biphenyls) and for nearly forty years they were considered to be the answer to all the problems associated with mineral oils. They were totally incombustible, did not form a sludge, had less tendency to emulsify with water, would withstand high temperatures, did not form explosive gases when subjected to arcing and had at least twice the permittivity of mineral oils. The higher permittivity, which is what determines the capacitance provided by a dielectric in a given capacitor, was similar in value to that of the solid insulating materials, with which they were used. For transformers in particular, this then provided a more uniform distribution of electric stresses than was obtained with mineral oil. In addition, the greater permittivity provided increased capacitance.

PCB is the mnemonic for a family of over 200 isomers of chlorinated biphenyl. The biphenyl molecule consists of two benzine rings connected at one point on each ring as shown in Figure 10.3.

There are then ten remaining sites where a hydrogen atom can be replaced by a radical such as chlorine. Strictly speaking the term polychlorinated should refer only to isomers having two or more chlorine atoms in the molecule, but is commonly used to refer to the biphenyl molecule with one to ten substitutions.

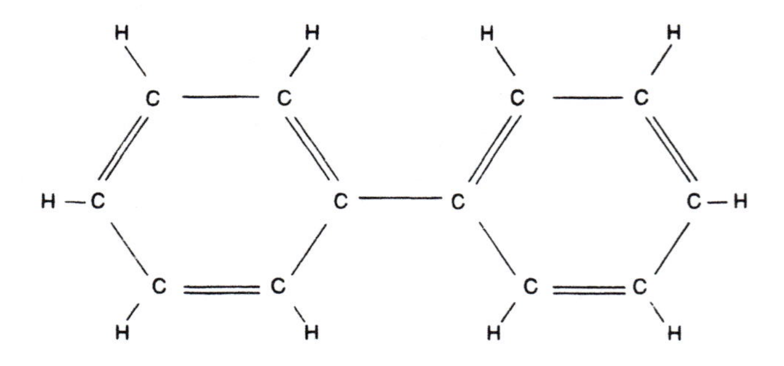

Figure 10.3 *The biphenyl molecule. Any or all of the hydrogen atoms can be replaced with chlorine to form polychlorinated biphenyl PCB. The durability of the substance increases with the number of hydrogen-to-chlorine substitutions in the molecule*

PCBs were discovered in the late nineteenth century and became widely used during the 1930s for transformers and capacitors. They were not considered suitable for oil immersed circuit-breakers as arcing produced decomposition products consisting partly of hydrochloric acid HCl. It was known that PCBs were highly toxic but it was not realised until about 1970 that the qualities that made them attractive to the electrical industry and for other purposes such as hydraulic systems, also made them persistent in the world environment. PCBs are no longer manufactured in the United Kingdom or in the United States. It is not known how much has already been released to the environment. As a result of accumulation in the food chain, the substance has been found in every organism analysed from the North and South Atlantic. It has been reported that in 1981 virtually the whole of the population of the USA had detectable levels in their bodies.

PCB enters via the lungs, the stomach and the skin and remains in the fatty tissues where it can cause chloracne, malfunction of the liver, reproductive effects and tumours. Its great resistance to the effects of heat means that the only satisfactory method of disposal at the present time is by high temperature incineration—above 800°C.

No polychlorinated biphenyls are dangerous while sealed into capacitors and transformers. The problem now is that there are millions of these older items still in service in factories, offices, hospitals, schools, shops, public buildings and houses and that they will all need to be replaced. PCB is a viscous but very penetrating oil that finds its way through apparently sound welds and seals. When a leaking capacitor is to be replaced or an old lighting fitting is to be scrapped, the PCB content is most likely to end up on the municipal land-fill site. The correct procedure is disposal in sealed drums to an authorised destructor.

10.3 Bibliography

HOLMES, L., *Lithium primary batteries—an expanding technology*, Electronics and Power, August 1980, pp. 658–660, Publ. IEE

MILNER, R., JUDSON, R., HARRISON, A.I., *Focus on d.c. supplies*, Electrical Times, 24 June 1983, pp. 7–10

BS 6132: 1993: *Code of practice for the safe operation of alkaline secondary cells and batteries*, Publ. BSI

BS 6133: 1995: *Code of practice for safe operation of lead/acid stationary cells and batteries*, Publ. BSI

VINCENT, C.A. *et al.*, *Modern batteries*, Publ. Edward Arnold, 1984

Installation and operation manual for nickel cadmium batteries, Publ. Alcad Limited, UK

RICE, M., *P.c.b.—filled capacitors—a little adds up to a lot*, Electrical Times, 4 Feb. 1983, p. 10

Chapter 11

Electric welding

There are two ways in which electricity is used for welding; by an arc and by resistance. Both depend on heat generated by an electric current.

11.1 Arc welding

With arc welding, two basic mechanisms are possible. The everyday method, as used for steel fabrications and general repairs and which can be undertaken by amateurs working in their garages and front and back yards, uses the heat from one end of an electric arc to melt the work pieces and the heat from the other end to melt a welding rod. Two metal parts can be fused together in this way with additional metal incorporated from the molten rod. This is technically known as manual metal arc welding, or MMA welding. The arc has a temperature of several thousand degrees centigrade and some of this heat is transferred to each of the two electrodes, that is to say, the welding electrode and the work piece.

To strike the arc a potential of about 80 volts is needed, but once the arc is established, a volt drop of only 20 to 40 V is required. Either a.c. or d.c. may be used and the supply can be from an engine-driven welding generator, with a falling characteristic, or from a mains-fed welding set consisting essentially of a step-down isolating transformer and a current limiter. The latter is normally termed a regulator and corresponds in its function to the ballast in the control gear for discharge lighting circuits. For welding purposes, the regulator should be controllable to provide a current which is appropriate to the work in hand.

The technique of MMA welding is closely concerned with the specification of the welding electrodes to be used for each application. A complex, internationally-agreed (ISO) coding system covers a vast range of electrodes having the following variables.:

- The material in the rod and the mechanical properties of the metal deposited in the weld
- The type of flux (normally provided by the coating on the electrode rod)
- The electrode efficiency (i.e. the percentage of the metal of the electrode which can be expected to form part of the weld)
- The welding angles or positions for which the rod is designed

- The recommended electrical conditions (e.g. polarity, if d.c. and the minimum open-circuit voltage)
- The level of hydrogen inclusion in the weld (diffused hydrogen in a weld renders it more subject to stress corrosion cracking).

In addition, there are a number of standard electrode diameters to suit the nature of the work and the available power supply. For instance, the commonly used 3·15 mm (⅛″) diameter rod needs 80 to 125 amps whereas a 6·3 mm rod would take 200–370 A.

The formulation of the coatings of electrodes is a technology in itself and plays an essential role in the production of the weld and in its final appearance and qualities.

The electrical characteristics of an arc differ from an ohmic resistance in several respects. Firstly, there is a more or less constant volt drop between each electrode and the arc, which is a conducting path of ionised gas. Secondly, the volt drop in the arc increases with its length so that, for a given supply voltage, if the welding rod is drawn away from the work piece, the arc will be broken and will have to be re-struck by dabbing the electrode into the molten weld pool. The column of ionised gas molecules is known as a plasma and contains both positive ions and their displaced electrons. The ions move relatively slowly in the electric field whereas the electrons, having far less mass, constitute most of the current flow. The fast-moving electrons collide with un-ionised gas molecules and knock out further electrons to contribute to the current and lower the effective resistance in the arc. Conversely, the arc resistance will increase as the current is reduced and, when it has fallen to between 5 and 10 amps, there will be insufficient electrons colliding with the gas molecules to create a conducting plasma of ionised gas. The arc will then be extinguished. Without an electric field, the recombination of ions and electrons is fairly rapid. Hence, with an alternating current it is more difficult to maintain an arc because it may not re-strike after each current-zero. To overcome this some inductance in the circuit is required. Current-zero is then displaced in time from voltage-zero and a continuous arc is more easily maintained.

Inductance can be provided by the voltage regulator. It means, however, that arc welding imposes a lagging low power factor load on the supply. This can be corrected by suitable p.f.-correction capacitors, but if these are left permanently across the welding transformer input terminals, there will be a corresponding leading low power factor whenever welding work is briefly interrupted. The load cycle can be improved by circuits which automatically switch in capacitance as the load is applied.

As mentioned, the supply characteristics need to have an open-circuit voltage of 80–100 V, falling to rather less than half this value on load. It is universal practice with manual welding to have the work piece at earth potential. Danger from shock is therefore chiefly with respect to contact with the welding electrode when there is no arc. The supply unit can be designed so that open-circuit voltage is reduced until a low resistance contact to earthed metal is made.

An alternative welding process uses a tungsten electrode which contributes no metal to the weld but merely acts as one pole of the arc. This, in conjunction with an inert gas fed round the work, enables various non-ferrous metals and special alloys to be welded. The process is known as tungsten inert gas, or TIG welding. Fabrications in aluminium are carried out by TIG welding with argon gas to prevent oxidation of the molten metal. With this protective atmosphere the process can be automated to give precisely controlled high quality joints for a wide variety of metals and gauges of sheet material. The arc is maintained by high frequency sparks at 10 to 20 kilovolts which keep the gas between the electrodes ionised. These ionising sparks are produced by a separate transformer from the main welding arc transformer. Although the sparking potential is injected across the welding electrodes, the mains transformer does not see this voltage as the high frequency is bypassed by a capacitor and blocked by an inductance, as shown in the circuit in Figure 11.1.

With a rectifier in the power circuit, very low current arcs can be sustained permitting metal foils down to 0·1 mm gauge to be joined.

The high-frequency sparks will generally be in the megahertz range and will accordingly be capable of generating significant electromagnetic interference. Suitable EMC precautions are necessary. These will include careful earthing of the equipment and the screening of the HV circuit.

Many other processes involving an electric arc are possible, for both welding and cutting. Most of these require programmed and automatically controlled systems.

11.2 Resistance welding

This method encompasses a variety of techniques such as spot welding, projection welding, stitch welding, seam welding, butt welding and flash butt welding. The process is a relatively simple one, requiring the passage of a high current at very low voltage through the parts to be welded so that sufficient heat is generated to fuse the two metal faces together. Pressure is first applied by the two electrodes which carry the current. At each welding point, the two work pieces are subjected to a pulse of current. The copper electrodes pressing the parts or sheets together are repeatedly conducting this high current pulse, so they need to be water cooled with the water channels as close as possible to the working faces of the electrodes. General purpose machines are normally operated with a foot pedal and pulse timer, but robotic welders are being increasingly used with dedicated software for mass production, as in the road vehicle industry.

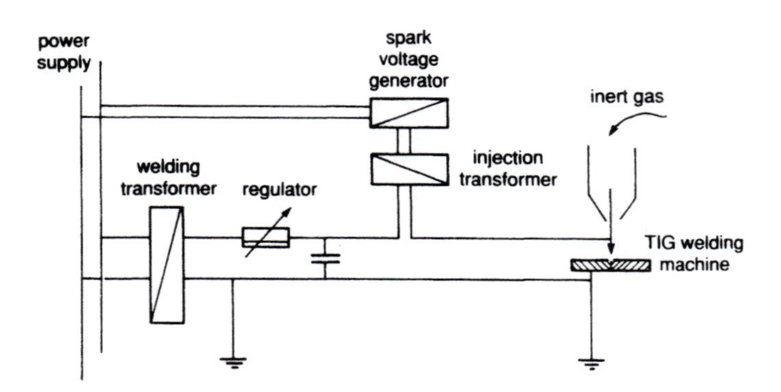

Figure 11.1 Schematic diagram of tungsten inert gas (TIG) welding circuit

For every application a specific controlled pressure, current and time is necessary for both manual and robotic welders. The electrodes will have small anvils to give a high localised pressure, or will consist of wheels profiled to the work concerned. Resistance welding is a method of fabrication similar to forging pieces of wrought iron together by hammering them on an anvil while hot—originally to improve the quality by hammering out the slag. Resistance welding has many more uses than this, including the joining of dissimilar metals.

Spot welding is a method of joining sheets of metal or other shapes by small local welds. It can be mechanised as a continuous process and applied to practically any structural metals and alloys. Intermittent pulses of current via wheels or rollers, which drive the work pieces along, produce what is termed stitch welding, or seam welding if the resulting joint is pressure tight. Another modification called projection welding uses parts whose mating faces have been embossed to give localised pressure zones.

Butt welding is applied to the machined edges or ends of rods and tubes. The end pressure, followed by the intensive current, causes some local swelling as the metal softens and fuses. With flash butt welding, a more highly developed process is used, the voltage between the two work faces being applied just before they are brought into contact. The result is a brief firework display as slag and some metal is expelled. With correct settings of pressure, current and timings, a uniform and relatively flush weld can be obtained.

11.3 Hazards associated with electric welding

One of the earliest types of welding electrode, dating back to 1911, was manufactured with a covering of blue asbestos (amphibole) yarn. This substance is now considered to be a serious risk to health and is proscribed in the United Kingdom under the Control of Asbestos at Work Regulations 1987, as amended by the Control of Asbestos at Work (Amendment) Regulations 1992.

Today there are still a number of hazards concerned with manual welding and especially with MMA welding, and these are summarised below.

11.3.1 Sparks

These can consist of sputtering incandescent metal and flux or slag.

When work is being undertaken on steel structures, these incendive particles can fall through open chequer plates and down the outside of tanks and on to other working locations. The need for general care and for protective headgear on construction sites is clear, as is the need to avoid any accumulation of combustible materials where welding is being carried out. The chief risk is from smouldering rubbish which becomes a fire after everyone has gone home.

11.3.2 Toxic fumes

Metal dust and other airborne substances can accompany welding processes. Some metals (such as beryllium) and their compounds are among the most poisonous substances known. An alloy of beryllium and copper can be used as a facing for electrodes in resistance welding. Persistent inhalation of welding fumes is capable of causing chronic damage to the lungs. There is also the possibility of asphyxiation when working in enclosed spaces and vessels.

11.3.3 Radiation

The electric arc produces intense UV radiation by virtue of its high temperature. Ultra-violet light is almost entirely absorbed by the transparent frontal part of the eyes; the cornea. Excessive exposure causes a painful, watery bloodshot condition known as arc eye. It is the same affliction as snow-blindness—also caused by excess ultra-violet light. At the other end of the visible spectrum, the heat from the arc also radiates in the infra-red. Undue exposure to IR light can promote the formation of cataract. Onlookers should not watch a welder at work without appropriate eye protection.

11.3.4 Hot surfaces

The metal in the neighbourhood of a weld will remain dangerously hot after it has ceased to glow. The end of the welding rod may cool rather more quickly but is capable of causing a painful burn.

11.3.5 Electric shock

The open-circuit voltage used for MMA welding is not likely to be lethal unless one is in a damp metal enclosure. In any case a severe shock can be received if the electrode is inserted into a live holder. There should always be a readily accessible means of making the electrode holder dead. Where possible the working position should be on an insulating mat and non-conducting boots should be worn for outdoor work.

11.3.6 Faulty earth returns

It is essential that the work piece has a sound and adequate earthed return connection back to the welding transformer or generator which must be connected to the site earthing system. It is dangerous practice to use existing protective conductors, or existing bonding conductors for this purpose. The return conductor should be reliably clamped to the work as near as practicable to the welding which is to be undertaken.

With fortuitous return paths the welding currents can produce local heating and perhaps some sparking at points of high resistance. In workshops and yards where welding is carried out at several places at once, each operator must be responsible for his or her own earth connection and return circuit to the welding set. Obviously, if welding is undertaken on slung work, the hoist chain should not be the sole means of completing the welding circuit.

11.3.7 Falls

A considerable proportion of site welding has to be done in awkward and hazardous locations and conditions. The fixing of working platforms and stairways on plant such as power station boilers, oil refineries and rigs, tank farms and so on, has to be done before there is any proper access. Normally insignificant shocks and burns can, in these circumstances, result in death by falling. Safety harnesses should be used.

11.4 Review

The welder must always protect him or herself with suitable personal protective equipment PPE as a defence against some of the above hazards. This equipment should include safety boots, apron, gauntlets, head, face and eye protection. It is also very advisable to keep ones overalls buttoned at the neck.

When working inside a vessel or any other confined space, several additional precautions are necessary. Depending on the circumstances, this will involve having an attached life-line and having a second person outside the enclosure ready to haul you out if you are in trouble. Plenum forced ventilation and/or breathing apparatus may be needed as well before work can proceed.

Two other recommended precautions are:

- Always disconnect the supply to the electrode holder before laying it aside
- Do not wear metallic items such as rings, watch straps, bracelets, hanging key clusters, or even lucky charms.

11.5 Bibliography

VILLIERS, A.E., *'Electrical Engineer' Reference Book*, 2nd Ed., Section 16, Geo. Newnes, London, 1946.

GRAY, T.G.F. and ANDREWS, D.R., *Electrical Engineer's Reference Book*, 14th Ed., Section 22, Butterworths, 1985.

Health and Safety Executive Guidance Note PM 64: *Electrical safety in arc welding*, HMSO, 1987.

Chapter 12

Lightning phenomena and protection

12.1 The nature of lightning

This title can be summarised by the letters ESD, for lightning with all its power and complex behaviour is no more than an electrostatic discharge. The awesome crack of a lightning stroke close by is no different in principle from the faint tick which accompanies a spark from the finger to the filing cabinet. Both are the shock waves from the sudden thermal expansion of air produced by a static spark.

As with other electrostatic phenomena, the charging processes in an electrical storm are not completely understood. It is evident that precipitation from a cloud, whether in the form of hail, rain or snow, does not create highly charged conditions. It is only when the precipitation takes place within the cloud itself that the situation becomes more interesting. Thunder clouds reach a height of several kilometres while their base may be less than a kilometre above the ground. They are thus several km deep. Such clouds can be produced by a cold wind driving under an extensive mass of warm moist air and so lifting it, the colder air being more dense. This occurs chiefly in the temperate zones, which are subject to winds of differing temperatures. In the tropical zones moist air may simply rise by local heating effects, causing a single cumulonimbus thunder cloud. In the polar regions, where there is virtually no warm moist air, there is hardly any lightning.

It is estimated that over the whole world there are up to 100 lightning strokes to the ground every second. For lightning protection design purposes the frequency of lightning storms is generally based on the number of strokes to the ground per km^2 per year. Empirical values for the United Kingdom are shown in Figure 12.1.

In other regions, such as parts of South America and in Madagascar, values of 140 flashes to the ground/km^2/yr are normal and in parts of equatorial Africa the figure is as high as 180.

The creation of static charges is generally the result of separation or division of insulated bodies and this also seems to be true in the charging of storm clouds. The upper part of a cumulonimbus cloud, being at an altitude of several kilometres, will be tens of degrees Centigrade below zero. The rising warm air will first have formed the cloud by the condensation of water vapour into microscopic droplets and as the air continues to rise the water vapour forms ice crystals. These water droplets and ice particles then coalesce to form rain and hail respectively. The

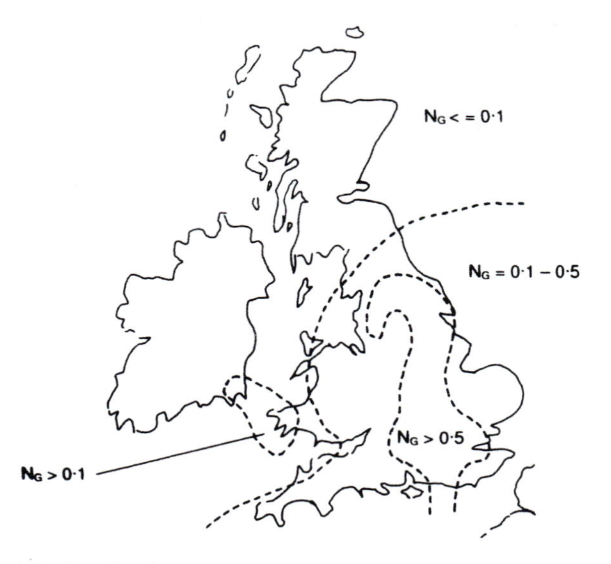

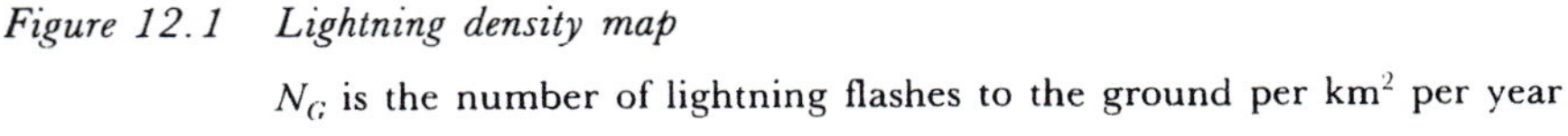
Figure 12.1 Lightning density map

N_G is the number of lightning flashes to the ground per km^2 per year

fierce upward draft in the centre of the cloud becomes progressively slower as it rises until the hail is able to fall back down into the cloud, passing through the raindrops. It is believed that the electric charging of the cloud is due to the hail breaking up the water droplets as it passes through a rain zone within the cloud.

It is generally agreed that the ice receives a positive charge and that the liquid phase acquires a negative charge. The warmer lower part of the cloud is therefore usually negative although about 5% of lightning strokes are from a positive charge at the cloud base.

The resulting ESD manifests itself in a number of ways. Most discharges take place within the cloud itself or within the cloud system. This is sometimes known as sheet lightning and is seen only as a flickering illumination of the overcast sky. Some discharges which do not reach the ground may be visible where they strike from the base of one cloud to another some distance away. These are termed air discharges. The usual visible lightning stroke is the discharge from a cloud to the ground, or a building or tree. This is the most destructive form and is discussed below in further detail. Before doing so, however, we should perhaps mention ball lightning which is one of the most mysterious and least understood of all natural phenomena.

It is always reported as a luminous sphere about the size of a football or slightly smaller. In most cases it is said to float along near the ground after a lightning stroke and then to 'explode' causing considerable damage. There are stories of ball lightning passing through the walls of a building and reappearing on the other side. In fact there are many anecdotes concerning ball lightning but seemingly no recorded or photographic evidence. It has been suggested that it perhaps represents a state of unstable equilibrium between positive and negative charges. Others have believed it to be a very highly charged suspension of liquid particles.

12.2 Development and characteristics of a lightning stroke to the ground

The first stage consists of the formation of a very high field strength at the cloud base. This intense potential gradient produces some ionisation of the air below the cloud with the consequent creation of a corona sheath. From the lower part of the corona, a brush discharge in the form of streamers will then extend progressively downwards in steps of about 50 m so that the conducting zone moves closer to the ground. This zone will be pierced by a downward spark of a few metres in diameter, termed a leader. Near the ground the leader penetrates the corona at a speed of about 300 km/s.

The concentration of charge on the boundary of the conducting zone is thus brought closer to the ground, increasing the field intensity and creating a counter upward charge from the earth and in particular from projections such as buildings and trees, as shown in Figure 12.2.

When a leader has brought a descending charge front low enough to break down the remaining air gap, brush discharges or streamers of opposite polarity are drawn upwards towards the descending charge front. A conducting path from the cloud to the ground is then complete and the main or return stroke takes place. This is the moment of discharge and is the first visible evidence of the process. We can say that the leader lowers the charge from the cloud down towards the ground and the return strokes then neutralise all the charge centres in the lower parts of the cloud. There may be several return strokes but they happen in such rapid succession that they usually appear as a single event. Each return stroke will move upwards several times faster than the downward speed of the leader and may carry a current of up to 400 000 amp. It is estimated that of the ground strokes

99% will be > 3 kA
50% will be > 28 kA
1% will be >200 kA

or in statistician's terms, the median value is 28 kA, the upper centile is 200 kA and the lower centile is 3 kA.

Because the corona sheath and the boundaries of a thunder cloud have no corners

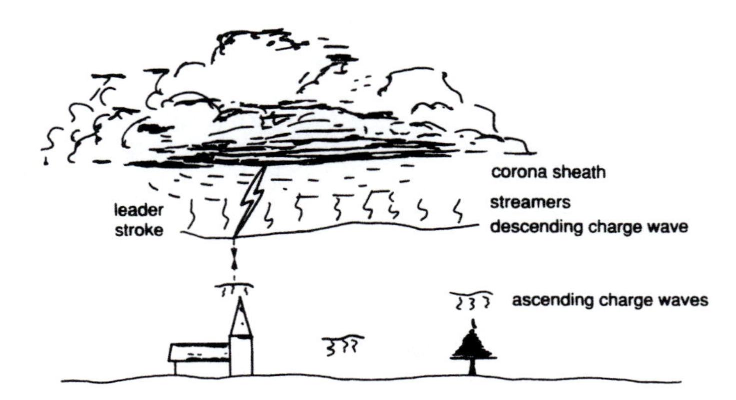

Figure 12.2 Electrostatic processes leading to a lightning flash

or small radius projections, a far higher potential can be present than on the surface of an angular solid object. By making certain assumptions as to the effective charge Q and capacitance C of a single storm cloud, or lightning cell, a value of the potential V of the cloud with respect to earth has been suggested, using the equation

$$Q = CV$$

where Q is in coulombs, C is in farads and V is in volts. For an assumed charge of 100 coulomb and a capacitance of 10^{-7} farad, this gives a value of 1000 MV for the potential.

Even if it is only one thousandth of this, the potential behind a lightning strike is still in the megaVolt range and its destructive power will be liberated whenever the path of the discharge meets an impediment, i.e. an electrical impedance. This can be likened to a bullet travelling through the air at high speed where very little energy is liberated until it hits something solid. When insulating or semi-conducting materials are subjected to a lightning strike, or find themselves in its path, the energy released will have the effect of an explosion at that point. Masonry can be blown off a building and obviously fires can be started.

The current in a typical lightning discharge will vary with time as shown in Figure 12.3. No two strikes will have exactly the same characteristics but the rate of rise of current will always be extremely steep. Empirically the maximum rates of rise are found to have a median value of 30 kA/μs, an upper centile of 200 kA/μs and a lower centile of 10 kA/μs. After the peak value has been reached, the current will decay approximately exponentially and reach half its peak value in some tens or hundreds of microseconds. The high rate of rise ensures that the first few microseconds of the discharge pulse will have a frequency spectrum which includes major components of several megahertz and harmonics of more than 100 megahertz.

With these very steep-fronted pulses, extremely high voltages will appear across inductances of only a few micro-henries and a lightning strike will see any bend or loop in its path as a high impedance which it will have no difficulty in jumping across. By the same token, capacitance will be seen as having a relatively low

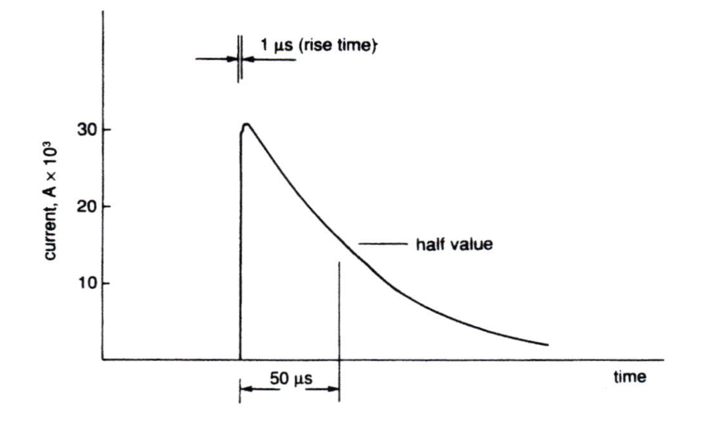

Figure 12.3 Time/current characteristic of a typical lightning stroke

impedance by the high frequency content of the rising current. This will also enable the discharge to be reflected in adjacent metallic objects.

A third errant characteristic of a lightning stroke to a building is known as side flashing. This is due to the sudden potential difference between metal parts resulting from the ohmic volt drop in the initial conducting path, as indicated in Figure 12.4.

In addition, the mechanical force produced by the high current itself can distort and dislodge metal conducting parts such as pipes and guttering. This is particularly relevant where two conducting paths run close together in parallel. Consider for example the arrangement shown in Figure 12.5.

The force between parallel conductors in air is given by

$$F = 2\mu_0 I_1 I_2 / x \text{ in newtons/metre run}$$

where μ_0 is the magnetic space constant which can be taken as 10^{-7}, I_1 is the current in one conductor in amps, I_2 is the current in the other conductor in amps and x is the distance between the conductors in metres.

The force is one of attraction between the conductors if the two currents are in the same direction and of repulsion when they are in opposite directions.

Assume a current with peak value of 100 kA produces a side flash to the waste pipe from the lightning conductor so that 20 kA flows down the pipe and 80 kA through the parallel conductor. The force drawing the two paths together will be

$$F = 2 \times 10^{-7} \times 20 \times 10^3 \times 80 \times 10^3 / 0{\cdot}1 \text{ N/m}$$

$$= 3200 \text{ N/m i.e. nearly a third of a tonne per metre run}$$

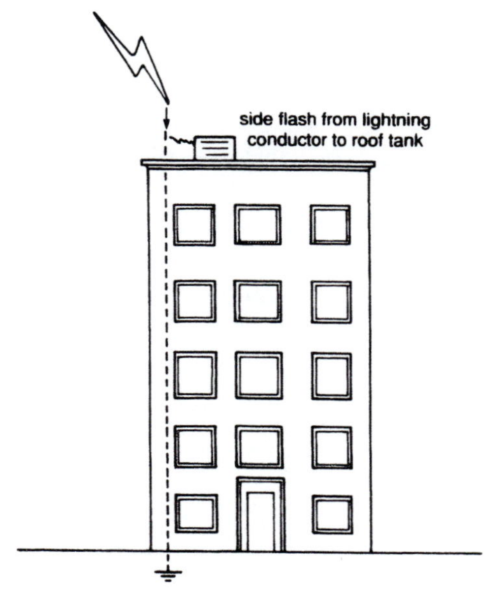

Figure 12.4 Illustration of side flash

At the moment of the lightning flash, the top of the conductor will assume a potential to earth of over a megaVolt while the water tank is at earth potential unless bonded to the lightning conductor system at roof level

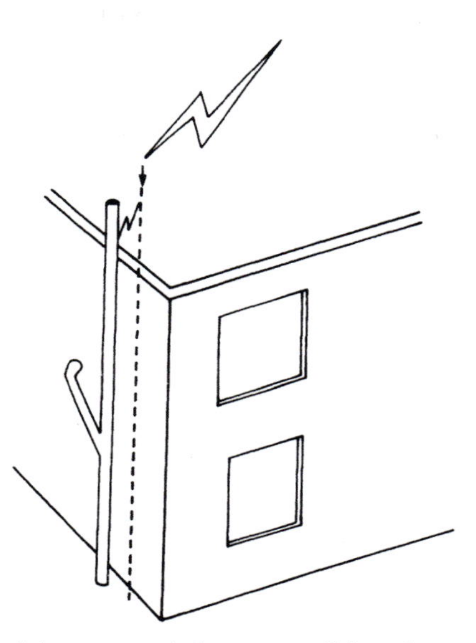

Figure 12.5 *Example of electromagnetic force on parallel conductors. The waste pipe, 10 cm from the lightning conductor receives part of the discharge current due to a side flash*

Similar mechanical forces on single conductors are produced by sudden changes of direction. Side flashes also release disruptive power from an acoustic shock wave caused by the explosive local increase in air temperature to several thousand degrees. The supersonic pressure pulse due to direct and side flashes can severely damage or displace building panels and roofing.

Lastly, but of increasing importance, we have the electromagnetic radiation from the pulse of high current which will bring about a disturbance at some distance. Lightning discharges at several kilometres distance are noticeable as a white noise on amplitude modulated (AM) radio receivers and nearby discharges can upset or corrupt computer and other electronic systems. Man-made sparks, such as from petrol engine spark plugs, can be suppressed, lightning discharges cannot and it is accordingly necessary to protect vulnerable electrical apparatus. The subject of electromagnetic interference is further discussed in Chapter 14.

12.3 Protection of buildings and services

The lightning conductor for the protection of buildings was invented by Benjamin Franklin (1706–1790), the son of a puritan soap and candle merchant who emigrated to Boston. All Franklin's considerable scientific researches were done in his spare time between international political and diplomatic work. Electricity was previously believed to consist of two fluids termed vitreous and resinous respectively. Franklin proposed a single fluid theory and coined the terms positive and negative electricity.

His famous investigation into the nature of lighting in which he was able to draw sparks from the end of the string of a kite while holding the string above the end, was regarded as convincing evidence that lightning was an electrical phenomenon. As he lived to report on the success of his experiment, it seems fairly certain that his sparks were simply from a charge on the kite itself and not from a storm cloud.

The effectiveness of a lightning conductor depends on a number of factors. Firstly it must be of uniformly low resistance. Secondly, loops must be avoided and changes in direction should be kept to a minimum and should not be sudden. Thirdly the conductor must be firmly secured to the building in order to withstand the possible electromagnetic forces. Lastly, the upper end of the conductor should be slightly above the highest point of the building it is protecting. The screening effect of a lightning conductor is reckoned to be within a cone whose half angle is 45°, as indicated in Figure 12.6.

Where a leader stroke comes down well to the side of a building, the return stroke may start from a corner of the structure some metres from the top. For this reason an additional conducting girdle should be placed round taller buildings a few metres below the parapet. The rule of thumb which assumes a protected zone under a 45° cone, should not be applied further away than 20 metres at the base. This is the same as saying that lightning conductors on buildings over 20 metres high will not protect others over 20 m away. Where buildings of various heights are clustered together an alternative recommended method for the determination of zones of protection is to use a notional sphere of 60 m radius, as shown in Figure 12.7.

As regards the cross-section of a lightning conductor, 50 mm^2 can be regarded as normal. It may seem surprising that a conductor of this size is sufficient for a current of perhaps 400 kA peak. The reason lies in the short duration of the stroke. When designing electric power wiring installations the temperature rise of cables and wiring needs to take into account the possible value of a short-circuit current and its duration before the protection is able to clear the fault. For this purpose the following simple formula is used:

$$I^2t = kS^2$$

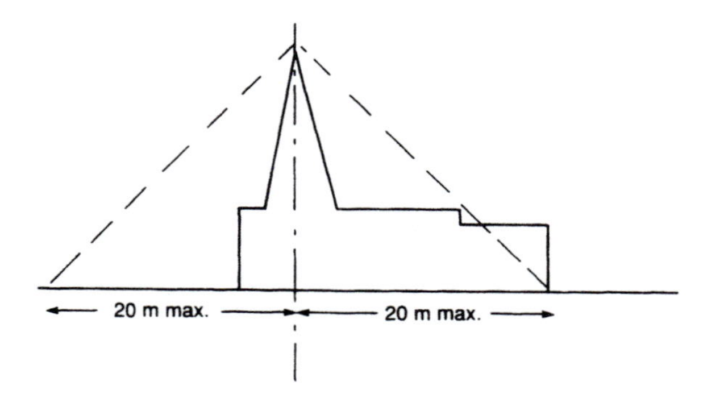

Figure 12.6 *Zone of protection provided by lightning conductor on the highest point of a building. Structures outside the cone with a 45° half angle need to be additionally protected*

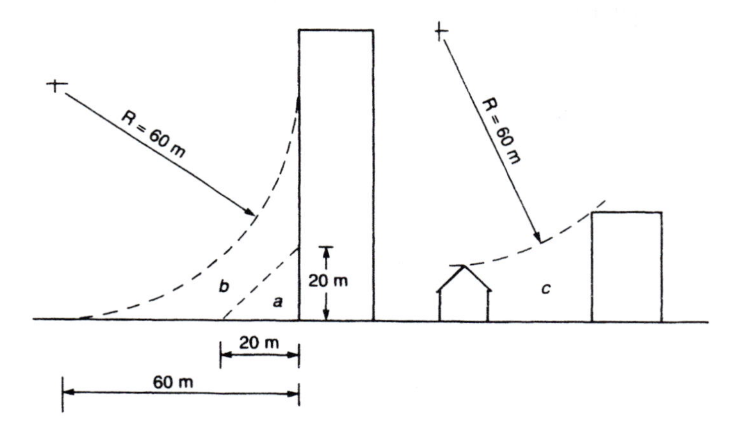

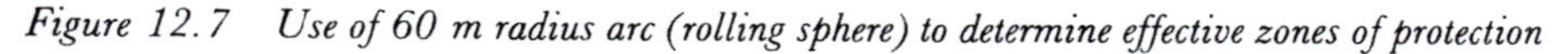
Figure 12.7 *Use of 60 m radius arc (rolling sphere) to determine effective zones of protection*

a Zone of protection given by a building up to 20 m high
b Zone of protection given by a building taller than 20 m
c Zone of protection between buildings

where I is the fault current in amperes, t is the duration in seconds, S is the cross-section of the conductor in mm^2 and k is a factor depending on the conductor material and the permissible temperature rise. For a copper conductor and a temperature rise of 160°C

$$k = 2 \times 10^4$$

For a lightning strike with a root mean square (rms) value of current of 100 kA, lasting for an equivalent period of 50 μs, the required conductor cross-section would be

$$S = I(t/k)^{1/2} = 100 \times 10^3 \,(50 \times 10^{-6}/2 \times 10^4)^{1/2} \text{ mm}^2$$
$$= 5 \text{ mm}^2$$

Hence, 50 mm^2 is more than sufficient from a thermal point of view and should be subject to a temperature rise of 160/100°C or 1·6°C in the above example.

It is preferable to attach lightning conductors to the outside of a building to avoid side flashing to internal metalwork and electrical apparatus as far as possible. The conductors should be reliably connected to structural and other metalwork for the same reason. This will include steel framing, concrete reinforcement (re-bars), handrails and metallic service pipework. Where the structure is metal-clad additional conductors to the ground may be unnecessary, but a flat roof of non-conducting material should be protected by a grid of conductors not more than 10 metres apart. External aerial support brackets or straps and rooftop air handling plant should also be directly connected to the lightning protection system.

12.4 Earthing of buildings and lightning conductors

The lower end of a lightning conductor has to be suitably taken into the ground.

This can be via concrete foundations containing adequate interconnected steel reinforcement, sheet piling, or by earthing electrodes. In most instances and with older buildings such as churches and brick chimneys it is necessary to install earthing electrodes. These can be in the form of rods, plates, mats or strips. If the soil is suitable for rods to be driven into the ground, they are more practical than plates etc. which always require excavation and reinstatement. The lower the resistance to earth the less the voltage elevation to be expected on the conductor during a lightning strike. Accordingly, it is normal practice to design the system so that the resistance to earth will not rise above 10 ohm. This value is something of a compromise between what is generally achievable and what is to be preferred. Even so, high soil resistivity can present problems in obtaining a value down to 10 ohm. The resistance is usually less at greater depths due to the presence of moisture. Electrodes of about 2 metres length should be used where possible as this depth will in most cases prevent seasonal variations. If bedrock is near the surface it may be necessary to drive a multiple array of electrodes connected to each down-conductor, or to install interconnected copper tape in back-filled trenches.

The diameter of a rod electrode is determined by the strength required to enable it to be hammered into the ground, 12·5 mm being a common size. If larger sections are used the cost increases as the square of the diameter, whereas the effective resistance is only marginally reduced.

An important consideration in the earthing arrangement of lightning conductors is the possible voltage gradient on the ground. In the vicinity of an electrode one can receive a shock from a lightning stroke as indicated in Figure 12.8. This shows person A at risk from the potential between points c and d (step potential), person B at risk from the potential between points a and b (touch potential) and person C at risk from the potential between points a′ and e (transferred potential). The voltage gradients are greatest near the electrode and decrease rapidly with distance. When cattle shelter under trees during a storm, they are subject to greater risk from step potential than a human would be since the distance between their hind and forelegs is greater than a person's stride and the voltage will be applied across the animals' cardiac region.

A practical method of reducing voltage gradients near an earthing electrode is to install the rods so that the tops are about 30 cm deep, with insulated connections. A typical termination within a hard-standing area is shown in Figure 12.9.

If the building to be protected has its own sub-station, the earthing of the lightning protection system and of the transformer should be connected together. Opinions differ as to whether the earthing should be combined with electronic system earths. To avoid interference from voltage spikes and surges on the mains supply, sensitive data handling circuits are regarded as requiring a so-called clean earth. A separate dedicated earthing system, physically remote from the lightning and power system electrodes is therefore frequently specified for new installations.

12.5 Protection of tank farms

Each tank should be equipped with at least one earthing stud near ground level. Each of these must be connected by a copper strap (e.g. 25 mm × 6 mm) to an

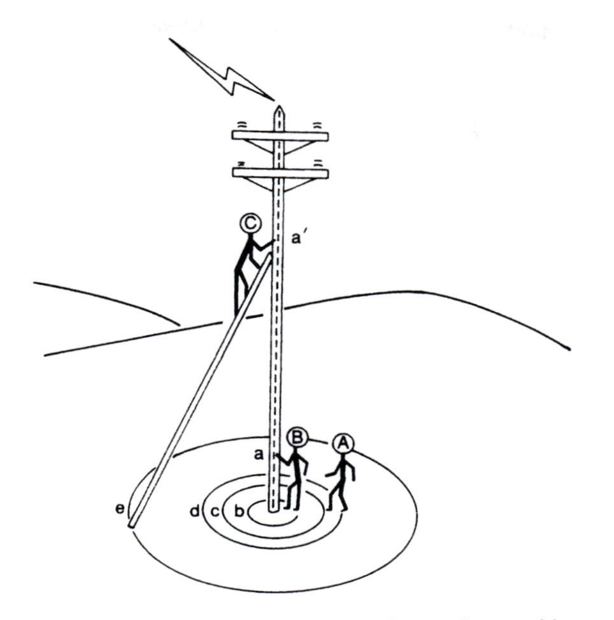

Figure 12.8 Effects of potential gradient on the ground around an earthing conductor during a lightning strike

A is subject to step potential c–d
B is subject to touch potential a–b
C is subject to transferred potential a'–e

adjacent earth rod driven through the foundation slab and bonded to the foundation reinforcement. All tanks on one slab should be bonded together.

All foundation slabs should be interconnected by these earth straps and, if there is a sub-station on the site, the whole earthing system should be connected to the main sub-station earth bar.

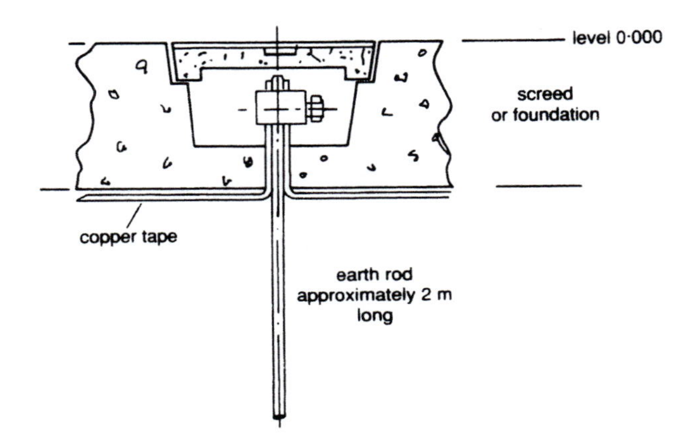

Figure 12.9 Earth rod top clamped to earth bonding tapes

All metalwork, including structural and access steelwork, metal pipework, deluge systems, bedplates, pipe stanchions, cable trays and any isolated metal items must be bonded to the earth system. Special earthing clamps are available for RSJs, re-bars, pipes etc. For bonding purposes, yellow/green insulated, stranded copper earthing cable, or bare copper strap can be used. Smaller items, such as cable trays, are usually bonded and earthed with 16 mm^2 cable, or larger sections where demanded by the prospective fault current of the power cables.

Earthing rods are typically of copper clad steel, 19 mm diameter by 2 metres long, and it is advisable to drive these in at not less than 4 equi-spaced points round each foundation slab. The tops of these should then be bonded to the general earthing system as indicated in Figure 12.9.

The earth for any associated control room containing computerised instruments will normally require two or more earth rods connected together but not connected directly to the main earthing system. These rods should preferably be driven outside the main earthed area so that power switching surges do not affect the process control computer.

12.6 Summary of lightning protection

- Where lightning conductors are used, they should be connected to the main power supply earthing system.
- Lightning conductors should have a resistance to earth of not more than 10 ohm. The electricity supply undertaking can usually advise on local soil conditions.
- Lightning conductors protect an area within a cone whose surface is 45° to the vertical—as shown in Figure 12.6.
- Lightning conductors should have as direct a route to earth as possible. Their earth rods should therefore be driven into the ground below, as close as possible to the earthing stud of the structure or to the down conductor.
- Reinforcing bars or RSJ pedestal bolts may be used to earth a lightning conductor, provided there are multi-crossing points in the foundation reinforcement ensuring electrical continuity to the general earthing system.

12.7 Statistical risks

It has been estimated that over 20% of computer failures are the result of lightning strikes. Some, but by no means all of these failures, can be prevented by protection of the building. When lightning strikes a building there may be damage to the structure causing injury by falling material. The building, however, does protect the occupants from direct electrocution by lightning and most of those killed and injured by lightning have been in the open. The annual risk of death by lightning in the United Kingdom is about one chance in two million. These odds are some 200 times longer than the risk of death on the road. Nevertheless, it implies that about 25 people are killed by lightning every year and probably more than half these will have been electrocuted by the potential gradient in the ground. Many others are 'struck' but survive after suffering various traumas including breathing arrest, paralysis of parts of the body, burns and shock.

12.8 Bibliography

BS 6651: 1992, *Code of practice for protection of structures against lightning*, Publ. BSI
Recommendations for the protection of structures against lightning, Publ. W.J. Furse & Co.Ltd., Nottingham (undated)
GOLDE, R.H., *Lightning Protection*, Publ. Edward Arnold, London, 1973

Chapter 13
Coping with static

Electrostatics is the study of electrical phenomena that take place when there are no moving charges. Static, the commonly used term for both charges and discharges, has many uses such as in printing and copying, dust precipitation, micro-electronics and in sprayed paint and powder coating. Electrostatic discharges (ESD) present many problems particularly with regard to ignition hazards, ranging from macro-scale disasters in grain silos and supertankers, to the explosion of the lungs of hospital patients while under anaesthetic. In addition, ESD costs the micro-electronics industry untold millions.

An awareness of static electricity is as old as recorded history, which is readily understandable since it is a natural phenomenon. The ancient Egyptians no doubt experienced static shocks when stroking their sacred cats and sailors have always been familiar with the brush discharge known as St Elmo's Fire appearing on the ends of their yard-arms. The Greeks knew that, when the fossil resin, amber was rubbed, it would attract other light bodies such as straw. In fact, the word electricity is from the word elektron, which is Greek for amber.

The term electricity was first used by William Gilbert (1544–1603) who carried out experiments on magnetism and static electricity. Figure 13.1 shows Gilbert demonstrating his work at the Court of Queen Elizabeth and it is surprising to realise that although it took place four centuries ago, the means by which static charges are produced have still not been fully explained.

13.1 Electric charges on solid surfaces

The most common way in which electric charges on solids are produced is by the separation of two surfaces in contact. The effect is increased if the separation involves sliding between the two surfaces. The results are well understood but the mechanism of the creation of charge is not.

It is believed that a transfer of charge occurs whenever two different surfaces touch, and that even though the two surfaces are initially at the same potential, at the point of separation there will be a potential difference between the surfaces of about one volt. The polarity of this will primarily depend on the nature of the respective materials.

If the surfaces are conducting, the charges are likely to leak away or be

Figure 13.1 *Dr. Gilbert showing his electrical experiments to Queen Elizabeth I and her Court*

This famous picture by A. Ackland Hunt, which hangs in Colchester Town Hall, was presented to the town, Gilbert's birthplace, by the Institution of Electrical Engineers in 1903 to commemorate the tercentenary of his death

neutralised by a relaxation flow of current through the areas of the two surfaces still in contact, or through the last point remaining in contact before separation.

If one of the surfaces is of a highly insulating nature, current cannot flow across it to escape and charge will remain in place distributed over the surface. The important factor here is that, although insulators cannot conduct electric current, they can retain an electric charge and hence have an electric potential. And if they are in a non-conducting medium such as dry air, the charges are trapped and remain in place. In extreme conditions, both positive and negative charges can remain in close proximity on the same surface.

Static potentials are usually of the order of several thousand volts. Assuming that, when an insulating surface is separated from another surface it is left with a potential of about one volt, how is a potential of several kiloVolts created?

Clearly, the two surfaces act as a form of capacitor. The capacitance C between two plates of area a and with a distance between the plates of d is given by

$$C = k\frac{a}{d}$$

and the voltage between the plates for a given charge Q will be

$$V = \frac{Q}{C}$$

V being in volts, Q in coulombs and C in farads. Hence C decreases as d in increased and since Q remains constant, being trapped on the non-conducting surface, V will increase with d.

The one volt at the start of separation is the potential difference corresponding to d = 0. The potential produced as d increases can therefore reach many kiloVolts even with quite small charges.

A more orthodox or 'correct' explanation can be derived by considering the law of conservation of energy. When the surfaces are pulled apart we assume that a positive charge on one face is balanced by a negative (or deficit) charge on the counterface. The two surfaces will therefore attract each other and work has to be done to separate them. This work is stored as potential energy in the well known form of $e = \frac{1}{2} CV^2$. This is analogous to stretching a spring when the work done is stored in the spring as potential energy. Since C decreases as d is increased and the stored energy e increases with d, the increase in the value of V^2 must be more than proportional to d.

All this of course begs the question of why there is a transfer of charge when the surfaces separate in the first place. It can perhaps be argued that, as d = 0 before separation of the surfaces, C is infinite, so random charges can remain in either surface without creating a potential difference. On separation, $d > 0$ and the random charges then produce an increasing voltage as d becomes greater. The generation of static charge by friction is more difficult to analyse but could possibly be explained by a similar conservation of energy argument assuming that individual charges are fixed in place on both surfaces. This becomes more credible when one takes into account the fact that to produce static charges by friction or rubbing, both surfaces need to be non-conducting.

So far we have considered static charges. The problems and hazards with static, however, are mainly to do with discharges. If a charged non-conducting surface is earthed at one point, the discharge will be restricted to the immediate vicinity of the earthing contact. The rest of the surface will retain its distributed charge. Depending on the surface conductivity, the charge on the more distant parts will slowly leak away through the earthed point. If the whole charge is to be quickly removed, the whole area must be wiped over with an earthed conductor such as a damp cloth. Alternatively, the surrounding atmosphere must be made conducting. One method is to ionise the air in the vicinity of the charged surface. Ionised air, being an electrical conductor, allows the charges to escape. Ionisation can be achieved by an array of sharp points held at a static potential of several kiloVolts. Raising the humidity of the air does not necessarily make it conducting, but above about 65% RH, the surface itself becomes more conducting.

Even conducting bodies can remain charged if they are kept insulated from earth. In this case though, when an earthed contact is applied to any point on the surface the whole charge on the body will immediately discharge to earth, so creating a fatter spark with far more energy. The human body has a low resistance in the context of static voltages of several kV and a discharge to a filing cabinet or radiator for instance can be quite unpleasant. We may wonder why the potential does not return to its original low value as d is reduced to zero. Here it is necessary to refer again to the equation

$$C = k\frac{a}{d}$$

and note that the shock is usually at the finger tip where, although $d = 0$, a is also very small. The capacitance to earth of the body consequently remains at about the same value as it was in the middle of the room and hence the potential to earth remains high. In principle, static shocks can be reduced by increasing the area of contact so that the capacitance is correspondingly increased. Thus, it is better to grasp the 'nettle' of the filing cabinet drawer handle than to test it gingerly with the knuckle.

At one time static shocks and potentials were widely used in attempts to cure various neuroses. At the beginning of the 20th century, in a treatment known as Static Sparks, the patient sitting on an insulated stool was connected to a high voltage generator such as a Wimshurst machine. An earthed conductor was then brought near enough to the patient to draw sparks. According to the literature of the time, this had a 'stimulating effect which is of service in hysteria'. One feels that the traditional slap across the face might have been equally effective and certainly would have been both quicker and cheaper. The therapy for headaches was to bring the earth electrode close but not close enough to cause a spark. This was called the Static Breeze or Souffle. The most visually interesting treatment would have been the Static Bath. No bath was involved but the patient was charged to about 50 000 volts. This caused the hair to stand on end and was said to be of help to insomniacs.

Only in exceptional circumstances are static shocks likely to be lethal or even injurious. The main danger with static is its ability to cause accidental ignition of flammable gases and dust clouds and of explosives. The minimum energy needed to ignite clouds of combustible dusts ranges from less than 2 mJ to above 5 J. Most explosive gases require between 0·2 mJ and 1 mJ when in their most explosive ratio with air. Carbon disulphide CS_2 and hydrogen H_2 are more sensitive and will ignite with a 0·02 mJ spark while some explosives can be detonated at energies down to 0·001 mJ. Most of the values are well below the energy released in typical static sparks.

The spark from the finger to the filing cabinet for example will normally be discharging a voltage of a few tens of kiloVolts. The capacitance of the body can vary quite widely, but taking an average value of 200 picoF, the value of energy in a discharge from say 10 kV will be given by

$$e = \tfrac{1}{2} CV^2 = \tfrac{1}{2} \times 200 \times 10^{-12} \times (10^4)^2 = 10 \text{ mJ}$$

This, as noted above, is more than enough to ignite an explosive gas.

In general there is no wholly successful way to prevent the formation of static on non-conducting surfaces and various methods have to be used to deal with it. Static causes adhesion of dust in the printing industry and in the weaving of synthetic textiles. Polymer film and paper will acquire a charge as they pass through rolls and where they are lifted away from conveyor belting. It is a particular problem where the sheet is wound on to a reel because the charge will remain on the outside of the reel and build up until the potential is great enough to cause a violent discharge to the frame of the machine. Plastic belting can also accumulate charge leading to noisy and frightening flash-overs.

Static discharges will occur when the surface reaches a certain critical density of coulombs/m^2. A unit charge of one coulomb represents a surplus (or deficiency) of about 6×10^{18} electrons. In the Système Internationale d'Unité (SI) nomenclature this can be written as 6 ae (6 atto-electrons). A coulomb is delivered by one ampere flowing for one second and one joule of energy (e) (i.e. one watt-second) is required or is generated when a charge of one coulomb is moved through an electric field with a potential difference of one volt.

The names of the units—coulomb, ampere, volt and joule—commemorating prominent scientists have been well chosen: the French physicist Charles Augustin Coulomb (1736–1806) determined the law governing the force between electric charges; the French physicist André-Marie Ampère (1775–1836) determined the law governing the force between currents; the Italian physicist Count Alessandro Volta (1745–1827) invented the electroscope, an instrument to measure electric charge and the work of the British physicist James Prescott Joule (1818–1889) included studies of the heating effect of electric current.

Coulomb's Law states that the force F between charges q and q′ separated by a distance d is given by

$$F = \frac{qq'}{kd^2}$$

For practical purposes the coulomb is an unduly large unit. Our filing cabinet shock for instance, with a voltage V of 10 kV and a capacitance C of 200 pF would be releasing a charge Q of

$$Q = C \times V = 200 \times 10^{-12} \times 10^{4} = 2 \text{ micro-coulombs.}$$

The cause of the discharge is the high field strength produced by concentration of charge and the proximity of another surface which is both conducting and at earth potential. For two flat surfaces a centimetre apart, the breakdown strength of air at atmospheric pressure is about 30 kV. That is to say, it will withstand a field strength of up to 30 kV/cm. Field strength is a vector and hence directional quantity which can be expressed as the force which the field exerts on a unit charge. It is then given in newtons/coulomb. Alternatively it can be defined as the loss of potential per unit distance in the vector direction, i.e. in volts/metre.

The potential V is a scalar quantity and can be defined as the work or energy taken or given in moving unit charge Q through the potential diffrence from V to V'. Thus, with static electricity we are really only concerned with potential differences so that

$$V - V' = \frac{e}{Q}$$

The potential in space is inversely proportional to the distance from a point charge. If the charge is distributed over a surface, this law will only be true at distances great enough to ignore the shape of the surface or charged body.

Field strength, being equal to the gradient of the potential in space, can be compared to a gravitational field where potential is analogous to height, charge is analogous to mass and field strength is analogous to the steepness of the slope. As can be seen from Figure 13.2, the field strength vectors will always be at right angles to equipotential lines or surfaces. Equipotential lines are shown for equal steps of voltage and it can be seen that they are closest together nearest to the charge and where the radius of the surface is least. Referring again to our gravitational model, the equipotential lines correspond to contour lines on a map. The slope of the ground is steepest where the lines are closest together and always at right angles to the contours.

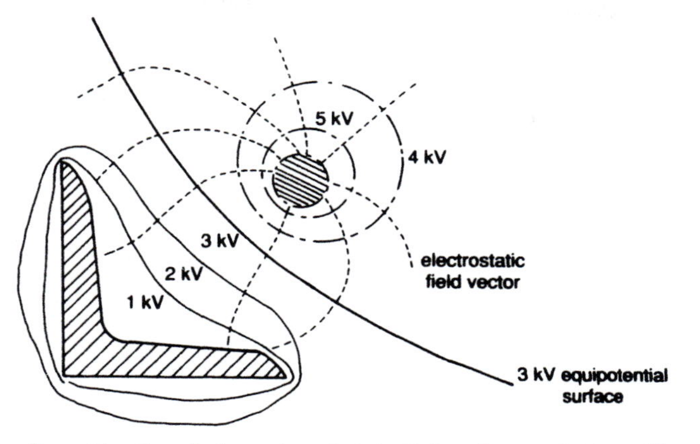

Figure 13.2 Example of angle iron at earth potential and a conductor with a potential of 6 kV, showing equipotential contours and electrostatic field vectors—two-dimensional representation

It can be shown both mathematically and physically that the maximum field strength will be in the proximity of the charged surface and will be greater where the surface is convex and less where it is concave. From this we note that the field within a charged vessel will be zero and that static discharges will occur first at spikes or sharp edges on the surface. The distribution of field strength is similarly indicated by the spacing between the equipotential lines in Figure 13.2.

13.1.1 Static-sensitive devices

An increasing difficulty with regard to electrostatic discharge (ESD) lies in its harmful effect on integrated circuits (ICs), particularly high-density micro-electronic chips containing multiple capacitors and field-effect transistors (FETs). Because these are rated at only a few volts, they will fail under a field generated by a static charge several centimetres away.

Solid insulation has a breakdown strength several times greater than air, as shown by Table 13.1. But although alumina, frequently used as a substrate, has a value of nearly 50 kV/mm, we recall that the voltage gradient or field strength is greatest at the electrode surfaces. The breakdown voltage is therefore a non-linear function of the thickness of the insulation.

To appreciate the vulnerability of IC devices, consider the working arrangement of bench assembly, where a person's nylon dust jacket has become charged to a potential of 20 kV. At a distance of 20 cm the equipotential shell in the air round the sleeve of the jacket may be only about 100 V and a micro-chip there will assume this potential to earth. Its internal packing density involves such small insulated distances between conducting parts that the breakdown strength is less than 100 V.

Although the energy, picked out of the air so to speak, will be minute, the IC components are correspondingly minute and can easily be wrecked by the resulting internal ESD. The most insidious aspect is the possible weakening of a device by the partial burning-out of interconnecting tracks so that it fails unexpectedly in service. Precision analogue amplifiers can also be degraded by ESD due to

Table 13.1 Electric strengths of some insulating materials

	Thick films (kV/mm)	Thin films (kV/mm)
Insulating oils	40 – 70	
Varnishes	20	
Polypropylene	30	
Polyamide (nylon/perlon)	17	65
Polyimide		280
PTFE		60
PVC		10 – 40
Glass	14	
Aluminium oxide (alumina)	48	
Air	3	

Average and approximate values are given since the breakdown voltage depends on several factors including time, temperature, waveform, thickness of the sample and shape of the electrodes.

sputtered metal. This may seem surprising in view of the tiny amount of energy liberated, but even if it is only a micro-joule, a 100 V discharge taking place in 5 nanoseconds will produce a current in the order of 2 amps.

The mean time between failures of solid state devices can be almost unlimited. Failures caused by ESD, however, make a nonsense of MTBF forecasting.

Damage to static-sensitive products caused by internal ESD can arise not only during the manufacturing stage but also during subsequent transportation and in service. During manufacture, special precautions are essential to minimise the generation of static. All unnecessary handling must be avoided. Earthed semi-conducting flooring and bench tops must be used and this also applies to other plastic surfaces and enclosures in the area. Leather-soled or other semi-conducting footwear should be worn and also non-synthetic textile clothing. The humidity of the air will reduce the ability of materials to retain a surface charge and this can be usefully applied for natural fibres and nylon which tend to absorb moisture from the atmosphere. The humidity required to affect more hydrophobic materials such as polypropylene will generally be too high for working comfort. An RH of 50% is said to be a good target figure.

For protection from ESD during storage and transportation, the methods of handling and packaging are vitally important. Semi-conducting or metallised wrapping materials should be used. Non-conducting anti-static materials are also available but these do not provide the benefit of a charge-free enclosure or 'Faraday cage' within the package.

Once in service, static-sensitive equipment needs to be housed in an earthed metallic enclosure.

Where semi-conducting materials are specified for preventing static from accumulating, a value of one megaohm to earth is normally regarded as satisfactory. This can be achieved either by incorporating a conducting powder such as carbon black in the mix, or by a suitable surface treatment. Ceramic insulators for example can be given a semi-conducting glaze. Footwear for use in electronic test laboratories, where live testing in an all-insulated environment is in operation, should have a resistance to the floor of not less than about one megohm in order to protect the wearer from a dangerous mains earth current. At the same time it must be semi-conducting to discharge static.

Problems due to static can be expected to grow as the size of micro-electronics and the spacing within integrated circuits continues to decrease.

13.1.2 Corona and brush discharges

Discharges from solid surfaces do not necessarily take the form of a spark from one body to another. When the field strength at the surface of a charged body exceeds the breakdown strength of the surrounding air, ionisation will take place. This enables some of the charge to leak away into the air since the ionisation allows current to flow. The discharge may be in the direction of another body at a different potential or may simply disperse as a leakage current into the atmosphere. The phenomenon is termed corona discharge and is accompanied by a white noise and sometimes by a slight glow round the charged body. The fizzing sound which is noticeable on EHV overhead lines in damp or foggy conditions is a form of corona discharge. In the absence of an adjacent earthed body the corona effect provides a limit to the voltage and charge density which can be sustained on a given surface.

As we have seen, the intensity of the field is increased where the radius of curvature of the surface is small and it will be greatest round a sharp point or edge. Statically charged conducting spikes will accordingly readily produce a corona discharge. In this case the ionisation will be present only in very close proximity to the point. The reason for this is provided by reference to Coulomb's Law which tells us that the field strength E at distance d from a point charge q is given by

$$E = \frac{q}{kd^2}$$

If the charge is distributed over the surface of a sphere, the lines of force radiating out from the surface would meet at the centre of the sphere if projected inwards. To determine the field intensity at a position outside the sphere we can therefore treat the total charge q on the sphere as though it were concentrated at the centre. The field intensity at a point d from the centre is therefore given by

$$E = \frac{q}{kd^2}$$

Hence we can see that, when the charge is concentrated on a spike, the field strength can be very great because d can be very small.

Due to the intensity of the field at charged conducting points, they are able to 'spray' charge on to adjacent conductors. Conversely, earthed spikes or needles can be used to draw off static charges from non-conducting surfaces without creating actual sparks. Thus, a row of sharp earthed metal points mounted very close to a fast-moving non-conducting surface can act as a static eliminator. The range can be increased by having a double row of spikes with a potential difference of about 12 kV between each row so that ionised air is produced irrespective of the static charge to be removed. This method is applied in many continuous processes, such as in the paper, plastic film and textile industries, where static is a nuisance because it makes dust adhere to the product. It can also be dangerous and noisy. Since static eliminators of this type can be designed so that they will not ignite flammable vapours, they are particularly important in colour printing where volatile (rapid drying) inks need to be used.

For the elimination of static from irregular surfaces, a combined blower and HV ionising electrode can be used. Alternatively, a suitable hand gun connected to an air hose and a 5 kV source of ionising potential permits static and dust to be removed from any local surface area.

Other static eliminators ionise the air in their immediate vicinity by means of radioactive isotopes. Alpha particle emitters incorporating radium or polonium can be very effective but require special health precautions. Beta emitters, such as thallium and tritium, operate at a lower intensity but are more easily screened and are less hazardous.

A corona discharge is a result of an electric field which is intense enough to ionise the air at one electrode while being insufficiently intense over the whole discharge path to permit a full spark. It can occur when the mean field strength between a point or sharp edge and a plane surface is as low as 5 kV/cm. As the corona approaches the plane surface a discharge spark can be produced. However, if the surface has very low conductivity, a large number of luminous branching threads may strike intermittently from the surface. Each of these threads is a small spark limited in energy by the amount of charge which can be fed into it by the non-conducting surface. This phenomenon is termed brush discharge because the total effect often has a brush-like appearance.

13.2 Electric charges on powders

Finely divided material in the form of powders and dusts are very subject to the accumulation of static. The charging effect is caused by a similar separation and sliding mechanism as with solid surfaces. Accordingly, non-conducting powders can become highly charged whenever they are poured, sieved, mixed, ground or blown. They can also transfer their charge to vessels and pipework.

The field intensity in the immediate vicinity of a particle will be determined by its size and shape and the surface charge density. The maximum charge density which can be retained on a surface is governed by the breakdown field strength of the surrounding air and, for both large areas and the micro-surfaces of finely divided and dispersed powders, an upper limit of about 10 $\mu C/m^2$ can be assumed for the charge density. As a powder becomes bulked or compacted, the individual charges will tend to combine and the mass of powder will then act as a very highly charged solid. Before compaction, the surface area of a given amount of powder is inversely proportional to the size of the particles. Hence, the finer the powder, the greater the total charge which can be sustained.

When a powder is in an earthed container, the charge will leak away at a rate dependent on the effective volume resistivity which will include the contact resistance between particles. True volume resistivity of a substance is the resistance across the opposite faces of a metre cube and is expressed in ohm-metres (Ωm). Alternatively, we can refer to its conductivity. Because it is the reciprocal of resistance, a unit of conductivity has been called a mho, but the SI unit is a siemens

(S) so that 1 MΩ = 1 μS and a volume resistivity of 10^{12}Ωm = 1 pS/m.

While the initial rate of loss of charge is proportional to the conductivity, the continuing rate will decrease and be proportional to the remaining potential. The decay will therefore be exponential so that the percentage decrease from any time t_1 to t_2 will be the same if $t_1 - t_2$ is the same. For an exponential decay there are only two parameters; the initial value and the time constant T which is the time to decay to zero if the initial rate were to have been maintained. It is called the time constant because the value of T (usually given in seconds or micro-seconds) is the same at every point on the curve, as shown in Figure 13.3.

For an exponential decay, the residual value of charge at time T will be 0·368 of the initial value. At 0·7 T it will be approximately half the initial value and at about seven times the time constant it will have reduced to 0·1% of the initial value. For radioactive material (whose radiation also decays exponentially) it is usual to refer to its half-life, i.e. 70% of its time constant. In the case of charged insulating substances, the rate of loss of charge is expressed by their respective relaxation times which are in fact simply the time constant T of their decay curves.

The relaxation time of a non-conductor is thus the time it takes to lose about two-thirds of its charge. It is not so important with solid substances but is a significant property of powders and liquids. Powders having a volume resistivity of 10^{12} Ωm will have a relaxation time of about 20 seconds. With such powders, discharges can occur within dust clouds and between dust clouds and the bulk powder. Incendive ESD can happen during filling and transfer processes and in

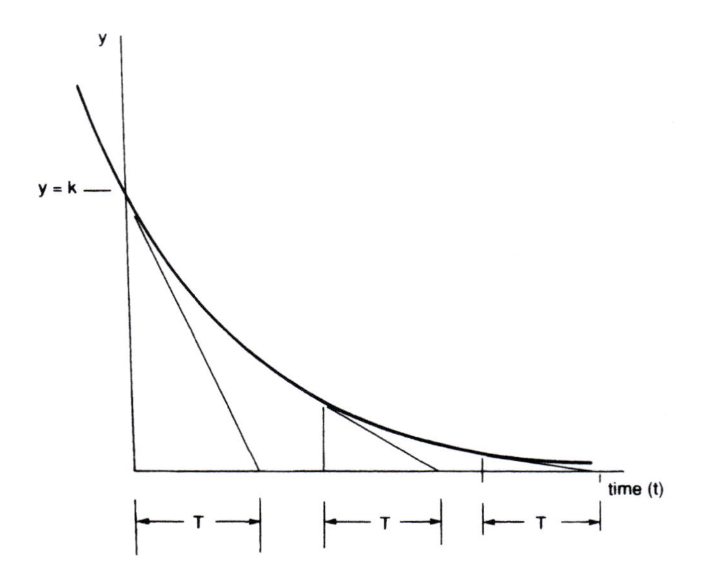

Figure 13.3 An exponential decay

The curve is given by the equation:

$$y = k\,e^{-t/T}$$

where k is the value of y at $t = 0$
e = 2·7183...
T is the time constant

the 1980s a number of major dust explosions in grain silos in N. America led to detailed studies of charge generation during silo filling and in similar processes.

When handling high resistivity powders it is necessary to earth all containers, especially if the powders are flammable or combustible. It should also be remembered that an ESD is more likely when some of the powder has been poured and has settled so that the charge has become concentrated—in the same way as the charge accumulates when a plastic film is coiled into a roll. For safety reasons, if low conductivity powders are to be poured from one container to another, the fixed container should be earthed and ones footwear and the flooring should be at least semi-conducting.

As mentioned above, some powders will not readily lose their charge even in an earthed container. On the other hand, an ESD from a bulk powder having a very high resistance will have limited power due to the low conductivity of the surface of the stored powder. Insulated pipes or ducting can become charged to a high potential when powders are blown through them. Similarly, jets of dusty air or of steam impinging on insulated metal can cause violent and dangerous discharges.

13.3 Electric charges in liquids

Oil and LPG companies and refineries and the chemical industry in general handle and store vast quantities of flammable liquids which are non-conducting and hence capable of retaining static charges. All through the 20th century these products have produced large scale disasters, often because a vessel exploding or rupturing in a tank farm or refinery can have a domino effect. On 28 July 1948 for example, a series of such explosions wrecked large parts of the Badische Anilin und Soda Fabrik plant at Ludwigshafen. The chemical complex extended for 10 km along the Rhine and the explosions—which continued at intervals for several hours—set fire to the town of Ludwigshafen and were so violent that there was severe blast damage in Mannheim on the opposite bank of the river. Over 200 people were killed and more than 1000 injured.

Static electricity in the chemical and petroleum industry is clearly an unwelcome and dangerous phenomenon which is difficult to avoid, but has to be treated with respect.

Charges in liquids are generated in the same manner as with solid materials, that is to say, when dissimilar materials are separated or slide past each other. Dangerous potentials are created in several well known ways the most important being

(*a*) Liquid flowing through a pipe
(*b*) Liquid flowing through a fine mesh filter
(*c*) Undissolved or immiscible particles moving within the liquid
(*d*) The presence of two different liquid phases
(*e*) Mist, spray or droplets moving through a gas or falling within a vessel.

With mechanism (*a*), there is a separation of charges between the flowing liquid and the wall of the vessel or pipe. When the fluid medium assumes one polarity, the wall of the container will be charged with the opposite polarity. If it is a metal pipe, the charge will flow away to earth and the remaining charge carried on by the liquid can be regarded as an electric current. When the speed of the liquid is great enough to become turbulent, the charge generation is increased and becomes far more than proportional to the rate of flow. Field strengths proportional to both the square and to the cube of the velocity have been quoted.

A vivid demonstration of mechanism (*a*) can be made by using a variable speed pump to circulate high resistance fluid through a section of metal pipe connected to earth via a motor-car spark-plug. The plug will spark at a very constant rate which increases as the flow is speeded up.

Mechanism (*b*) is more complex and the charging rate of liquid flowing through a filter is difficult to predict. It is known that the finer the filtration, the greater the charge generation and it is possible for a filtration process to produce far higher currents than a pipeline on its own. We speak here of currents although the subject is still a matter of static electricity, but carried pick-a-back on a moving non-conductor. Other factors affecting the current generation are the state of the filter elements and the rate of flow, the electrical resistance and the ionic state of the fluid, none of which permits simple prognosis of the charging effect.

Charging by mechanism (*c*) is likewise not amenable to calculation but is nevertheless a very important cause of electrification. It is known from practical experience that the presence of water in a high resistance liquid will significantly increase its accumulation of charge. This can also take place without pumping, by water dispersions settling through the product. Oil companies regard the presence of water as highly dangerous, many incidents having been attributed to this cause. As a rule of thumb, products known to contain any dispersed water should not be pumped faster than 1 m/s.

Solids moving through a liquid will also cause static charging, e.g. during stirring or mixing. When there is water or sediment at the bottom of a tank, it will inevitably be disturbed during filling, leading to additional charge generation. Here also a maximum filling speed of 1 m/s is sensible.

Mechanism (*d*) has a similar effect to mechanism (*c*).

Mechanism (*e*) is of special importance when filling tanks and other vessels. The inlet pipe should be at low level or, if at high level, a fill pipe reaching nearly to the tank bottom should be used. Should the liquid be allowed to enter from the top, droplets and splashes will be produced. These may be charged as they leave the filling pipe and may pick up further charge density on the way down. Where the atmosphere within the tank is explosive, this so-called splash filling is obviously a dangerous procedure.

An object in the path of particles flowing through air or vapour will be fed with electric charge and, if insulated, may then discharge by a spark to earth. Steam is a true vapour but, when released, water condenses to form droplets, so creating a visible fog. If escaping steam is directed on to an unearthed part, charge can be transferred from the droplets giving rise to sparking potentials. Fires resulting from escaping steam may be due to carelessness or to ignorance of the hazard.

A steam leak on the roof of a gas analyser house caused a spark from an unearthed piece of metal on the roof which set fire to the building. In another instance, during an inspection of an oil refinery in the Middle East it was noticed that an escape of hydrogen through a small leak in a high pressure joint was being temporarily diluted by playing a wisp of steam at the leak from a plastic hose. No accident occurred before the plant was able to be shut down and repaired, but a breeze of clean, dry air from a bonded metal pipe would have offered less chance of an ignition.

13.3.1 Relaxation times and anti-static additives

In the context of static electricity the ability of a liquid to retain a charge is its most important property. This is governed by its relaxation time which is synonymous with the time constant T of its exponential decay and is approximately 1·44 times the half-value time. As one might expect, T is a function of the volume resistance of a liquid and is in fact proportional to it. The relaxation time is important because charge is generated at a certain rate and if it can leak away as fast as it is generated no dangerous potentials will arise.

It is usual to relate T to the conductivity rather than the resistance when considering flammable liquids. For most petrochemical liquids the conductivity will be in the region of pico-siemens per metre (pS/m). A value of 1 pS/m is termed a conducting unit (c-u). In terms of resistance, 1 c-u is thus $10^{12}\ \Omega$ m. One c-u gives a relaxation half-life of 12 s.

Natural petroleum (crude oil) contains a very wide palette of substances, from the lightest distillates down to asphalt. The lighter, or most volatile ends, will have c-u values between 0·1 and 10, the lower conductivity corresponding to a relaxation half-life of about 2 minutes. The conductivity of the crude itself may range from 1000 up to 100 000 c-u depending on its source and its half-value time will usually be below 10 ms. Accordingly, crude oil does not present a hazard as regards static although, due to the presence of the lighter fractions, it is always capable of generating a flammable atmosphere. Water also has no ability to retain static charges because its half-value time is about a micro-second even though, in the context of electric currents, pure water can be considered as non-conducting.

There is always a risk of ignition if a pump or filter discharges low conductivity liquid into a tank containing an explosive mixture through a short line giving the liquid inadequate time to relax. In many instances danger from static can be avoided by allowing the charged medium to rest before entering the vessel. A dwell time in an earthed line or buffer tank of 4 times the half-life will reduce the charge to 2^{-4} of the initial value—a reduction of nearly 94%. Alternatively, the charging of an incoming liquid can be lowered by using an inlet section of line of larger diameter, to reduce the speed of flow.

Metallic salts, being electrolytes, are conducting and proprietory mixtures of salts are available as additives to raise the conductivity of a product. Very small quantities are needed and these anti-static additives can often be used without affecting the quality of the product. In practice the resulting benefit is that pumping rates can then be safely increased.

Static charges will usually cease to be a problem if the conductivity is raised to 50 c-u, but higher values may be required for pumping rapidly through filters—as in the case of aircraft fuelling.

When in very small concentrations, even as low as one gram in 1000 tonne of liquid, some impurities can increase the charging effect, whereas for the improvement of conductivity to a safe level, over one gram per tonne of anti-static additive is needed. The type and dosage of anti-static additives can therefore be fairly critical.

13.3.2 Other methods of protection

A short relaxation time does not give a safeguard against all eventualities. For example, a vessel with a residue of crude oil in the bottom will probably contain an explosive atmosphere and, if water is introduced through a manhole in the top, the droplets can become charged to a sparking potential as they fall. Tank explosions have occurred when water has been added in this way from a canvas hose in order to raise the ullage to an outlet pipe level. Marine vessels have also exploded during tank washing with water cannon. After three supertanker explosions in 1969, while some empty tanks were being cleaned with high pressure sea water, extensive technical investigations were undertaken. The ignitions were all considered to be due to spray and packets of falling water, known as water-slugs, which between them produced static potentials capable of forming incendive sparks.

The explosions required the presence of oxygen. Air normally consists of about 21% oxygen and 78% nitrogen. The nitrogen acts as a buffer and reduces the force of the explosion. With pure oxygen, the explosion would be vastly more violent and would require only about 1% of the normal ignition energy. If the proportion of oxygen is further reduced, a level is quickly reached where no ignition can take place. Protection by this means is termed inerting. Nowadays, marine cargo tank washing is generally only undertaken after the tanks have been inerted from the ship's diesel engine exhaust. On fixed installations, the vapour space of a storage vessel can be kept in a non-explosive state by reducing the percentage of oxygen by introducing an inert gas such as nitrogen or a compatible flammable gas. Petrol for instance can be gas-blanketed with LPG or natural gas.

A full tank is reasonably safe from explosion even if it is statically charged since there is no space for an explosive atmosphere. A floating-roof tank or a container with an earthed floating cover is therefore safer than a fixed roof tank, but any enclosed space above a floating cover needs to be suitably ventilated. With very large storage tanks a floating roof is often the preferred solution because, after applying protection by inerting or gas blanketing, it will take several days before the atmosphere within the tank is uniformly mixed and during this period part of the space can remain explosive.

Another approach which can be applied in certain cases is to increase the conductivity of the vapour space by means of radioactive sources within the upper part of the tank. This is only a partial solution since the vapour remains explosive.

13.4 Electric charges in gases

The charging processes experienced with solids and liquids does not arise with gases. However, jets and streams of gas containing particles, either solid or liquid will allow the particles to generate charge from, or impart charge to, other items of equipment. The release of CO_2 or of LPG will cause considerable cooling leading to the formation of CO_2 snow and LPG mist which can both be dangerous from a static point of view.

Where compressed air is used for cleaning, the disturbed dust and spray present a possible hazard and all nearby metal objects should be suitably earthed.

13.5 Ignition of explosive gases

Although a particle-free explosive gas will not produce a static charge, it can be readily ignited by a very small or low energy ESD such as one may see in dry weather from a woollen blanket or silk nightdress. During the seven years between 1947 and 1954 there were 36 reported hospital accidents caused by the explosion of anaesthetic gases. The patients were killed in three of these incidents and many others have been injured by them. In preparation rooms, X-ray rooms and operating theatres the risk of explosion and the resulting hazard is not confined to the patient since an exploding rib cage can act as a fragmentation bomb, injuring medical staff in attendance. The majority of these ignitions were caused by static electricity.

A Ministry of Health working party which studied the problem made a number of recommendations:

(1) Cotton, linen or viscose rayon bedding and clothing should be used in place of wool and other synthetic textiles
(2) The chassis and enclosures of all medical equipment should be of metal or of a material having suitable anti-static properties so that all apparatus is electrically continuous throughout and with a conductive path to the floor
(3) All surfaces on which movable objects can be placed should be free from paint or other insulating finish
(4) Theatre floors should not be waxed.

It is clearly essential that all parts of anaesthetic apparatus should be suitably bonded. Plastic and rubber items and components such as wheel tyres, tubing, bellows, draw sheets, aprons, gloves and footwear must have anti-static properties. That means they must be at least semi-conducting with an upper limit of resistance of 100 megohm.

At one time the floors of many theatre suites were effectively non-conducting and it had long been considered necessary for the floors of X-ray rooms to be fully insulating because of the danger of electric shock from the high potentials used. The above-mentioned report recommended that in these circumstances, before explosive anaesthetics were used, the floor area should first be dampened. If the floor had a non-hygroscopic surface and the effect of damping the floor would be only temporary, a moist sheet should be put down over an area sufficient to ensure that a person approaching any part of the anaesthetic equipment would have to step on the damp area before touching the apparatus.

The working party on anaesthetic explosions was very aware of the benefit of a high relative humidity in reduction of risks of static discharge. Even though the atmosphere in surgical areas is usually fairly warm and dry, it was suggested that the damp atmosphere prevalent in the British Isles had prevented the problem from reaching anything like the magnitude it had—at that time—attained in America. With this in mind, the report suggested that the theatre sister or some other member of staff should always inform the anaesthetist whenever the relative humidity fell below 50%, but did not indicate what action the anaesthetist was expected to take. Generally there is no difficulty in finding a non-explosive alternative gas. There are, however, some medical cases and procedures for which the only wholly satisfactory anaesthetic is the explosive gas ether.

13.6 Electrostatic painting and finishing

The painting or coating of surfaces by a high pressure spray is a labour-saving technique which has no doubt been used for centuries. The development of miniature circuits able to generate potentials in the kiloVolt range has now allowed spray finishing to be assisted by electrostatic attraction with relatively simple equipment.

The hazards to be considered when operating with a hand-held spray gun are firstly, the possibility of high voltage electric shock and secondly, if a flammable atmosphere is involved, the danger of fire or explosion. When the work is carried out in a spray booth or within an enclosed space, there is also a respiratory hazard. Some materials and solvents may in addition cause skin irritation and rashes.

The normal process uses an air gun with a concentrically arranged electrode at very high voltage within the nozzle. The sprayed material may be a mist of atomised paint, a powder or flock. The electrode imparts a charge to the substance as it leaves the gun so that, after being expelled it is drawn to the workpiece by electrostatic attraction. For powder coating, the product will generally be subsequently heat-cured to produce a coherent finish. PVC-coated articles made from spot-welded steel wire—hanging baskets and kitchen plate racks for example—can be very cheaply produced by immersing the preheated wire assembly into the thermoplastic PVC powder. Other materials, such as metal powders, require a sprayed process. Flock finishes are usually blown on to a surface already treated with adhesive which can be slightly conducting. In this way a plastic base material can be maintained at a uniform earth potential.

Not only the workpiece but also the operator and the casing of the gun must always be at earth potential. Hooks and the whole of the suspension system for the work in process therefore have to be kept free from accumulations of insulating sprayed material. There must also be good contact between the operator and the gun which thus has to remain conducting over a substantial surface area of its pistol grip handle. The operator's gauntlets and footwear must be anti-static, but build up of paint is sometimes difficult to avoid and it may be preferable to use a glove with the palm cut out.

Electrostatic spraying equipment for use with flammable substances must be designed so that any spark produced will have insufficient energy to cause an ignition and explosion. The safety of the design in this respect can be reliably proven only by suitable practical tests. Hand-held spray guns for sale within the European market are accordingly type-tested in a representative explosive atmosphere. During the test, an earthed electrode is brought up to the sample gun with its high voltage source turned on, but its atomising air supply turned off in order to increase the chance of an ignition. Any discharge which the gun is capable of producing then takes place in an atmosphere which is not diluted and made less explosive by the atomising air.

Harmonised European standards require paint spray guns to be type-tested in this way in a specified propane/air mixture. Designs for which a prototype has passed this test are then marked 0·24 mJ, indicating their maximum discharge energy and their suitability for use with flammable paints and solvents. Spray guns for powders and flocks are type-tested in a specified methane/air mixture and approved designs are then marked 5 mJ. If flock is to be sprayed on to a surface in the presence of an adhesive which can form a flammable atmosphere, 0·24 mJ equipment has to be used.

Another, more recently published Euronorm, addresses the electric shock hazards of hand-held spraying equipment for non-flammable materials. With a potential of tens of kiloVolts, the internal capacitance contributing to a discharge becomes critically important as well as the power that can be delivered by the high voltage circuit. This standard specifies the following limiting values:

- A maximum capacitance between HV parts and earthed part of 2000 pF
- The maximum continuous current which can be drawn from the high voltage circuit is 2 mA
- The maximum energy which can be discharged from the high voltage circuit is 350 mJ
- A minimum value of the current limiting resistor in the spray gun of 10 kohm per kVolt of the maximum voltage which can be generated by the HV source.

Increasingly, electrostatic spray installations are being automated, particularly where many identical items have to be coated. The software for the required manoeuvres of the gun can be readily established by storing data generated by a skilled operator while manually spray-coating a sample workpiece. The instructions can then be fed to a robot as required.

13.7 Bibliography

FOORD, T.R., *Static electrification*, Electrical Review, 14 Feb. 1969, pp. 231–233

HEALTH AND SAFETY EXECUTIVE occasional paper OP5, *Electrostatic ignition*, 1982, Publ. HMSO

BS 5958: 1991 *Code of practice for control of undesirable static electricity*, Publ. BSI. Part 1: General considerations. Part 2: Recommendations for particular industrial situations

CAMPDEN, K., *Electrostatics—fields of interest*, IEE Review, Oct. 1991, pp. 335–338. Publ .IEE

BELL, K., and PINKERTON, A.S., *The fatal spark*, IEE Review, Feb. 1992, pp. 51–53, Publ. IEE

HMSO: 1956, Ministry of Health, Report of a working party on *anaesthetic explosions*

Chapter 14

Electromagnetic radiation

The more one thinks about this subject the more puzzling it becomes. Work at a distance, with nothing between the transmitter and the receiver is as mind-boggling as the square root of minus one. Electromagnetic radiation concerns not only electrical apparatus. The warmth from a coal fire as much as that from an electric fire and the light from a candle as much as that from a torch are all electromagnetic radiation. It is only the frequencies which are different from radio transmission, but whatever the frequency or distance, the speed of propagation through space is the same. One of the most fundamental constants of the universe is thus the speed of light.

An electric charge per se produces an electrostatic field but no magnetic field. When the charge moves, as for example in an electric current, a magnetic field is produced round the current vector, as indicated in Figure 14.1.

In absolute space, movement is only relative. Hence, although a stationary charge has no magnetic field, anything moving with respect to the charge will be moving in a magnetic field. The magnetic field round a constant current vector is unchanging and no power is radiated. If the current changes, the magnetic field changes with it and electromagnetic radiation is transmitted through space. Thus radiation requires not just movement of an electric charge, but a variation of the current vector. Continuous radiation is obtained when the current is alternating, so giving a sinusoidal acceleration of charge.

In 1820, Hans Christian Oersted (1777-1851), the professor of physics at the university of Copenhagen, accidentally discovered while lecturing that an electric current deflects a magnetic needle. In 1829, self-induction was also accidentally discovered by Joseph Henry (1797-1878) who later became a professor and director of the Smithsonian Institution in Washington. The relationship between electricity, magnetism and induction was discovered and studied independently by Michael Faraday (1791-1878) who became a professor of chemistry at the Royal Institution, London. Faraday also found that light is affected by magnetism.

Electromagnetic radiation was first demonstrated in 1888 by the German physicist Heinrich Rudolf Hertz (1857-1894) after James Clerk Maxwell (1831-1879), who became the first professor of experimental physics at Cambridge, had proved mathematically in 1864 that any point in space with a varying electric or magnetic field, will transmit that variation as a radiated wave travelling with

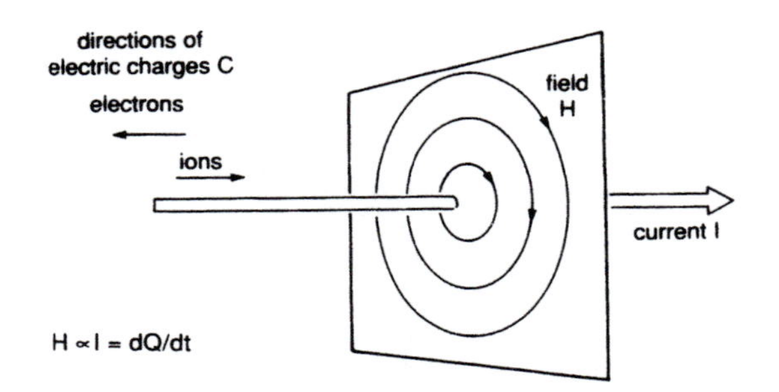

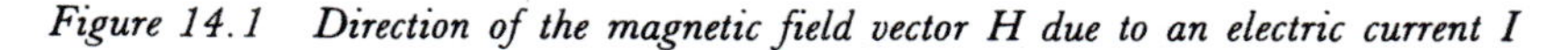
Figure 14.1 Direction of the magnetic field vector H due to an electric current I

The direction of H is taken as the direction in which a N-seeking magnetic pole would wish to travel. The direction of I corresponds to the direction of flow of positive charges in a conductor

the speed of light and deduced that light itself is propagated by the same electromagnetic waves.

It is now known that light and radio waves are two rather narrow bands in the whole electromagnetic spectrum which extends from the alternating current frequencies used by the electrical supply industry (ESI) to the highest frequencies known, namely cosmic rays. The various frequency bands are shown in Figure 14.2.

It will be seen that the range goes up as high as 10^{24} cycles/s (hertz). From 10^{0} to 10^{24} Hz is about 80 octaves whereas visible light, from violet at one end of the spectrum to red at the other, is less than one octave. Although the earth receives radiation at other frequencies, it is only the visible part of the frequency range that is able to reach the ground without undue attenuation. The eye has developed accordingly so that it can see only those frequencies which illuminate the ground.

Electromagnetic propagation through space can be likened to the wave motion which moves along a rope when it is vibrated at one end. The wave or ripple produced travels at a constant speed which is unaffected by the rate of vibration. The wavelength is therefore inversely proportional to the frequency. The amplitude decays very little with distance travelled because there is not a great deal of energy dissipated or released as the wave moves along the rope. Electromagnetic waves travelling outwards through empty space will diverge to a greater or less extent and will therefore be reduced in strength as the square of the distance travelled. However, no energy will be given up en route and in principle the radiation will continue for ever. The phrase 'for ever' is a cliché embracing the two immortal concepts of eternity and infinity, neither of which is comprehensible. The theory of evolution could argue that this is because an understanding of eternity and infinity is unnecessary for our survival and prosperity—in the same way as one is insensitive to electromagnetic frequencies which do not reach the ground.

The theory of electromagnetic radiation however is well understood, Maxwell having translated Faraday's perspicacious ideas of lines of force into the exact language of mathematics. From Figure 14.1 we see that a current is always surrounded by a magnetic field. The total magnetomotive force (mmf) round a conductor is proportional to the current and the magnetic field strength H is the

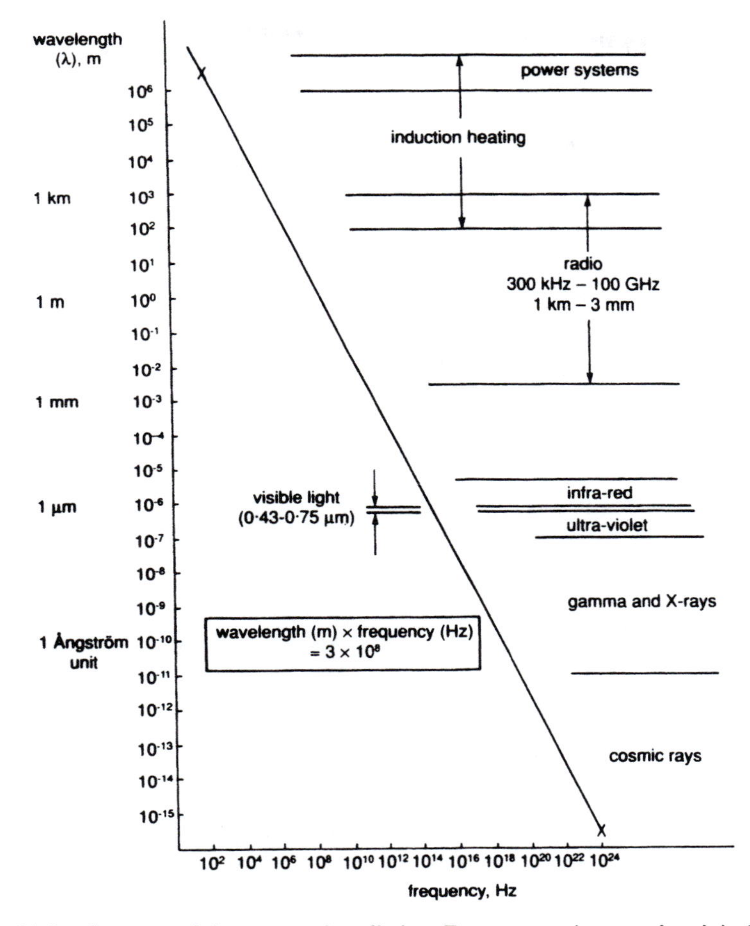

Figure 14.2 Spectrum of electromagnetic radiation. Frequency against wavelength in free space

total mmf round the magnetic path divided by its length. The cause and effect roles of electromotive force (emf) or E and magnetic flux density, which in free space is proportional to H, can be reversed as shown in Figure 14.3.

Here emf is proportional to the rate of change of H (dH/dt) in the same way as H is proportional to current or rate of change of charge (dQ/dt) in Figure 14.1. Hence, we can see that the radiated field strength is proportional to charge acceleration. The vector H is always at right angles to the electrical field strength vector E and both are at right angles to the direction of propagation. Thus, electromagnetic radiation is in the form of a natural rectilinear co-ordinate system in Euclidian space. Maxwell produced four so-called field equations which succinctly define the interdependent properties of electricity and magnetism in space. In the words of the physicist Max Born, these four simple formulae show wonderful symmetry and he adds:

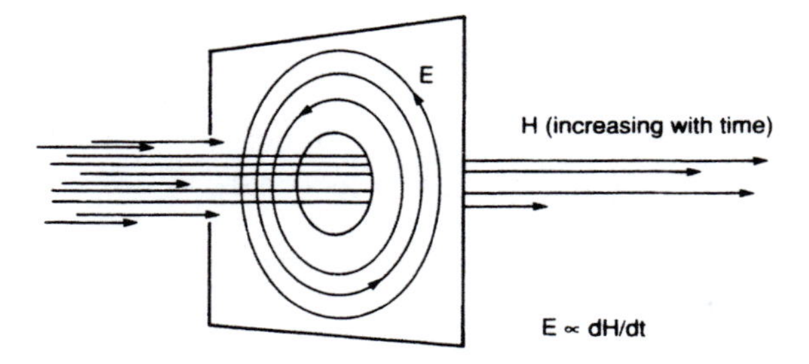

Figure 14.3 Direction of electromotive force vector E in relation to an increasing magnetic field H

The direction of H is taken as the direction in which a N-seeking pole would wish to travel. The direction of E is taken as the direction in which a positive charge would wish to travel

> 'Formal agreement of this kind is by no means a matter of indifference. It exhibits the underlying simplicity of phenomena in nature, which remain hidden from direct perception owing to the limitation of our senses and reveals itself only to our analytical faculty.'

The equations incorporate two constants which are fundamental to electromagnetic radiation. Free space has minute but measurable electrical and magnetic properties. These are, respectively, its capacitance per metre or permittivity, denoted by epsilon-zero (ϵ_0) and measured in farad/m, and its inductance per metre or permeability, denoted by mu-zero (μ_0) and measured in henry/m. They behave in a manner analogous to springs at right angles to each other and the rate of propagation in vacuo is equal to $(\mu_0\epsilon_0)^{-1/2}$. Both μ_0 and ϵ_0 are extremely small values and give a speed of approximately 300 000 km/s. Numerous experiments and measurements have shown that this is also the speed of light, c.

Diagrammatically, the transmission of energy in vacuo can be shown as in Figure 14.4.

A varying field H produces a linking emf E which causes a charging current to flow into the capacitance of free space. This so-called displacement current then produces a linking magnetic field H, the flux being determined by the inductance of space and the chain-like process continues with finite velocity. Figure 14.4 is

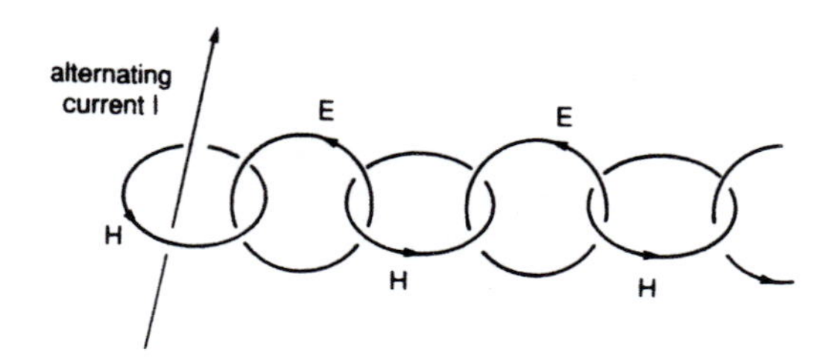

Figure 14.4 Simplified representation of electromagnetic radiation. When I varies sinusoidally, H varies sinusoidally in phase and E varies sinusoidally 90° out of phase

obviously a very simplified diagram as the radiation actually extends in all directions.

Insulating materials have higher values of permittivity ϵ, also termed the dielectric constant, kappa (κ), so that their rates of propagation are correspondingly slower than c.

Summarising the phenomenon of electromagnetic radiation, we can say that when a current changes, the surrounding magnetic lines of force are subjected to a change propagated into space at the speed of light c and this is accompanied by a simultaneous emf, whether or not there is a conductor enclosing the magnetic lines of flux.

14.1 Circuits which produce and accept radiation

As we have seen continuous radiation requires oscillation. A resonant electrical circuit is normally associated with radio transmission since this permits large swings of current with minimum energy loss in the circuit. Various forms of current oscillation are possible. The most simple system is by a spark discharge from an energised circuit. The sudden release of stored energy in the capacitance and inductance which all circuits possess to a greater or lesser extent can produce a high frequency response which rapidly decays exponentially. Radiation was first detected (by Hertz) in this way. Emission of continuous waves requires a constantly excited resonant circuit connected to an aerial. For optimum performance its impedance and natural frequency should be respectively matched and tuned to the impedance and frequency of the resonant driving circuit. There are many types of oscillator circuits and an indefinite number of aerial configurations. All of these rely on the magnetic storage of inductances and the charge storage of capacitances.

The efficiency of an oscillating system can be defined as the energy stored within the system at resonance divided by the energy per cycle to sustain the amplitude of oscillation. The ratio of the total stored energy to the energy absorbed per cycle is termed the Q factor. Systems with a high Q will resonate freely and will have a longer time constant when the excitation is removed than systems with a low Q; low values of Q correspond to highly damped circuits.

As a mechanical analogy we may consider a large church bell weighing say half a tonne. Once the bell is swinging, it requires a small amount of work each swing to keep it going. This is equal to the energy dissipated per cycle due to the internal damping of the system plus the acoustic energy emitted by the bell. The energy of the bell while swinging, varies sinusoidally between kinetic and potential energy and is far greater than the topping up work needed per pull. When the bell-ringer stops, the energy in the bell decays exponentially due to the energy absorbed each cycle. The greater the damping factor the quicker the bell will cease ringing. In the absence of damping, an elastic system would go on vibrating at its resonant frequency indefinitely. Conversely, if it were continuously excited at its resonant frequency the amplitude would increase until it flew to pieces. All systems have some damping, however, and this determines the amplitude of resonant vibration.

Excitation of a high Q resonant circuit can generate high and even dangerous voltages by a process of resonance amplification. These voltages will swing sinusoidally between inductive and capacitive components in a similar manner to the cyclic exchange of kinetic and potential energy in the church bell.

A resonant electrical circuit requires inductance L and capacitance C. The impedance Z of inductance increases with the value of L and with frequency f, so that

$$Z = 2\pi fL$$

The impedance of a capacitance decreases with the value of C and decreases with frequency, so that

$$Z = (2\pi fC)^{-1}$$

These impedances are in phase opposition and so cancel out when they are equal, i.e. when

$$2\pi f = (LC)^{-½}$$

For a hypothetical series circuit with no resistance (where Q is infinite), as in Figure 14.5, there would be no impedance if energised at resonant frequency. Circuits of this type have what is termed acceptor resonance. For a parallel circuit, as in Figure 14.6, the impedance at resonant frequency would be infinite but so would the voltages and currents in the L and C components. Circuits of this type have what is termed rejector resonance.

An interesting point to note is that since the resonant frequency is governed by the product of inductance and capacitance, a circuit with components having a high capacitance and low inductance can have the same resonant frequency as one with a high inductance and low capacitance. Of particular practical importance

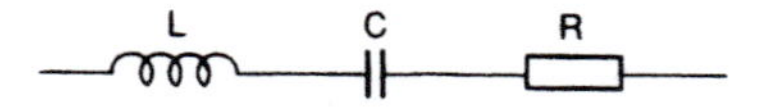

Figure 14.5 Circuit with inductance L and capacitance C in series

When frequency $= \dfrac{1}{2\pi(LC)^{½}}$

the impedance is equal to the circuit resistance R. A theoretical circuit without any resistance would have zero impedance

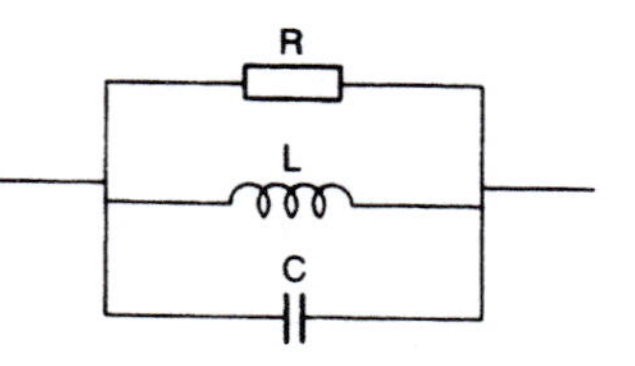

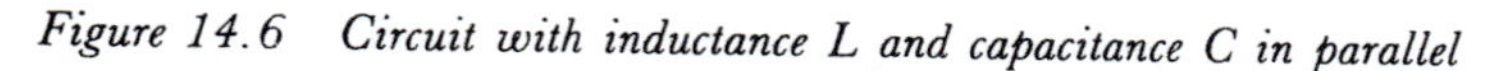

Figure 14.6 Circuit with inductance L and capacitance C in parallel

When frequency $= \dfrac{1}{2\pi(LC)^{½}}$

the impedance $= R$. A theoretical circuit without any resistance would have infinite impedance

is the fact that lower values of the product LC correspond to the higher frequencies. Thus, the higher clock rates of modern electronics and computers go hand in hand with miniaturisation.

14.2 Electromagnetic interference

Intentional radiation, as in broadcasting, radio communication, radar etc., is by means of a high frequency signal-generating circuit whose output is connected to an aerial (now commonly termed an antenna) which is designed to have specific characteristics. Its size will be a function of the wavelength or wave band concerned and its shape will depend upon whether it is to be directional or isotropic. In practice, no aerial can be fully isotropic, but it can be directional and have a field of view which is predominantly horizontal—in the manner of a stationary lighthouse beam. The benefit obtained is known as the gain of the aerial and is usually compared numerically to that of a simple dipole. The gain and other electromagnetic properties of an aerial will be the same whether it is acting as a receiver or a transmitter. Also, if an object such as a pole or line receives radiated energy, it will to a greater or lesser extent, then act as a transmitter reradiating the received signal. This property is utilised in the design of most short-wave aerials in order to improve their gain in a particular direction, e.g. the Yagi antenna.

Transmitting aerials may need to handle high voltages and often high powers. Fixed receiving aerials are usually both tuned and directional, but mobile or portable aerials are often a telescopic rod which is neither tuned nor directional.

In the early days of wireless broadcasting, the term interference was associated with fading and distortion of long distance signals. This is usually caused by multi-path transmission. If one path is different in length from another by an odd number of half wavelengths, the two signals will be in antiphase and will tend to cancel out, whereas the reception will be reinforced if the signals arrive in phase. Fading is generally experienced due to changes in the height of the ionosphere which alters the path length of the refracted signal, as indicated in Figure 14.7.

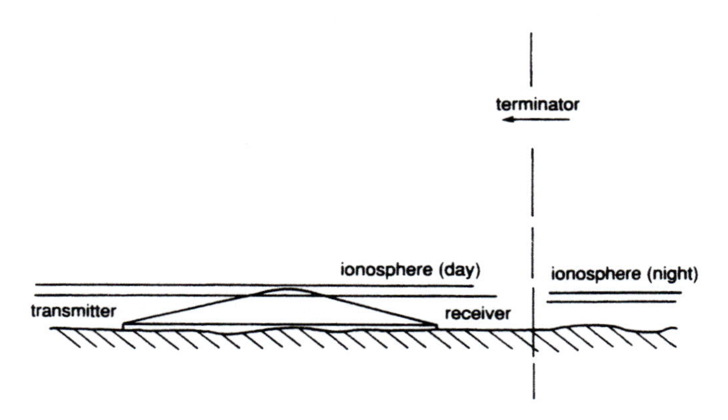

Figure 14.7 Direct path of radio transmission and indirect path refracted by ionised layer of the atmosphere. The height of the ionosphere is greater in sunlight than at night

Today the term interference normally refers to man-made congestion in the air from electromagnetic radiation. It is manifest in numerous ways: firstly, there is always the possibility of frequency trespass, where a signal is broadcast on a frequency already allocated to another transmitter in the area, or a broadcast signal is too powerful so that its range of reception overlaps the territory of another transmitter on the same frequency, or a transmitter is using a band width that is too wide so that it encroaches into an adjacent frequency band.

Secondly, there is interference from electrical apparatus and machines. Those which spark or arc such as igniters, commutator motors, petrol engines, welders, arc furnaces, surgical diathermy equipment, discharge lamps and signs, electric railways, trams and trolley buses are all particularly prone to cause interference by radiated energy. Because the radiation in these instances is unintended it will have random and wide ranges of frequency. That is to say, the radiation will have many harmonics, some of which can be received and amplified by apparatus dedicated to one specific frequency. Radiation due to sparking is also usually in short bursts of discharged energy transmitting far higher power than could be sustained continuously.

In addition there is radiated interference from natural causes such as lightning and varying radiation from the sun which can disturb the ionisation of the atmosphere. Electromagnetic pulse (EMP) from a nuclear explosion can also cause radio-interference over several million square kilometres.

For domestic radio and television receivers radiated interference can be a nuisance. More serious is interference in communication, data transfer and process control circuits, especially in the field of technological medicine. A typically sensitive area concerns cardiac pacemakers. These operate at very low levels of power and may transmit signals from outside the body to a receiver attached to the heart wall. Transmission of digital data over air links can readily be corrupted by spurious radiated pulses and, depending on the importance of the information, various countermeasures have to be applied. These include additional error-recovery signals, known as parity checks, redundant circuits and rejection of unrecognised instructions or data.

Directional aerials can differentiate to some degree between wanted and unwanted or interference signals. Further dedication to the correct signal can be obtained by polarisation of the transmission so that the field strength vector E is oscillating in only one transverse plane, with a corresponding orientation of the receiving aerial. The method corresponds to the use of polarised spectacles to exclude unwanted light. Such aerial characteristics are important for domestic TV receivers as spurious signals can affect the quality of both the picture and the sound. On the screen they may cause white or black speckles and if the interference is due to the television signal itself arriving by an indirect route, having been reflected or reradiated from a nearby object, a second or ghost image will be produced. Large moving objects in the vicinity, especially aircraft, cause transitory signal variations known as flutter. Other problems are sometimes due to radio and TV sets themselves which can adversely affect reception close by because their circuits contain oscillators and time bases which can radiate via their aerials or mains leads.

Persistent interference can sometimes be located by means of a portable radio with its own ferrite aerial.

As shown in Figure 14.8 the ferrite rod concentrates the flux along its length.

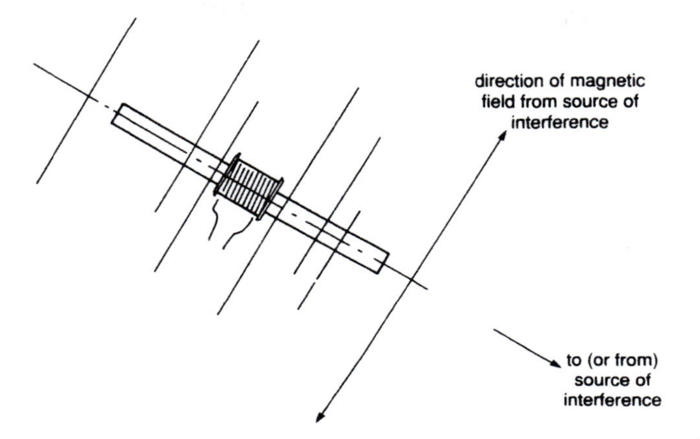

Figure 14.8 Ferrite aerial rotated to angle of minimum strength of interference signal

Remembering that the magnetic vector H is at right angles to the direction of propagation, it can be seen that, when the spurious signal is at minimum strength, the rod will be pointing towards or away from the source.

With ghosting of TV pictures the extra length of the ghost signal path can be calculated from the displacement of the second image if the line scan speed is known. For instance, if the ghost image position on the screen represents a delay of 4 μs, that signal will have travelled 4×10^{-6} c metres further than the direct signal (where c is the speed of light = 300 000 km/s), i.e. 1·2 km.

It is perhaps relevant to consider here which frequencies of electromagnetic radiation can cause interference in electric circuits. Although electromagnetic radiation can be at virtually any frequency, cosmic rays being up to some 10^{24} Hz, it is only the lower end of the spectrum, i.e. up to about 10^{11} Hz that produces radio waves as such and it is only radio waves which can be generated by or converted to electric currents. In other words, what we mean by radio waves are those frequencies of electromagnetic radiation which can be interfaced with electric circuits. The reason for the upper frequency limit is one of scale. Electromagnetic radiation has no electric current flowing since there are no electrons or any other charged particles in Maxwell's displacement current in vacuo. It is merely a fluctuating magnetic field H with its induced emf E. To this extent, the term 'displacement current' is an abstract mathematical concept. Currents in metallic circuits, however, consist primarily of electrons in motion. Their random thermal jostling has an overall drift—which is the current—in the direction of increasing positive potential, since electrons have a negative charge. Their mean rate of drift is only a few mm/s, which is in the order of a kilometre per day. A mechanical analogy is a very slowly moving tray of very rapidly scurrying ants. If each ant represents an electron, a current of one amp requires a migration of $6 \cdot 3 \times 10^{18}$ ants/s past a given line.

The important point is that, at the upper end of the radio spectrum, the electron drift will be reversing twice every 10^{-11} s, during which time the drift towards the positive potential will be about 10^{-10} mm, or less than a millionth of a micron. Above this frequency the electron oscillation reduces to molecular dimensions and we are in the infra-red and optical part of the spectrum. Normal aerial coupling between radiated fields and electric circuits is then no longer feasible.

14.2.1 Protection against interference

The most satisfactory way to deal with problems of interference is to tackle them at the source, or as the doctor would say, prophylaxis is better than therapy. Suppression at the source, however, is normally beyond one's control and various measures have to be applied to exclude unwanted signals. These measures can take a number of forms including the use of

- A directional aerial
- An aerial location selected for the best signal-to-interference ratio
- Circuits and systems with good signal discrimination and spurious signal rejection
- Screened leads for amplifier inputs and for all low level signal lines
- Screening of the apparatus itself
- A pre-amplifier close to the aerial or to the instrument probe, as the case may be
- A radio frequency filter on the mains lead to the equipment
- Careful and suitable earthing of the apparatus.

Where the interference is radiated at a single frequency or band of frequencies affecting the function of the apparatus, a rejection or filter circuit can be connected into the signal line. Troublesome individual frequencies can be dealt with by using a filter having a high impedance to that particular frequency. These are sometimes termed notch filters from the shape of their admittance characteristic.

Many types of apparatus, especially in the field of instrumentation have high gain amplifiers whose input signal is at a very low level of power and is received either via an aerial or by a screened cable. The screening, which is usually in the form of a braid or metallised plastic tape prevents an external field from generating a voltage on the cable core or cores so that stray signals are not fed to the amplifying circuits. This is a direct application of Faraday's cage, within which he personally demonstrated that an electrically charged field could not exist. The screening of electrical equipment is either by the metal housing or, if the casing is of insulating material, it may be lined with metal foil or a sprayed metal coating.

For a number of reasons, total exclusion of an external field is not generally a practical proposition and the degree of attenuation within the screened zone is an important consideration in design and installation work. It is commonly specified as the ratio of external field power P_1 to the internal field power P_2. The unit of attenuation is the bel and is numerically given as $\log_{10}(P_1/P_2)$. Thus, if the internal field power P_2 is $0{\cdot}0004$ of P_1, the attenuation is

$$\log_{10}(P_1/P_2) \text{ bel} = \log_{10}(2500) = 3{\cdot}4 \text{ bel}$$

In practice the decibel unit rather than the bel is always used and is written as dB. So in the above example, one would say that the screening (which is called shielding in American English) reduces radiated interference by 34 dB. As a rule of thumb, 3 dB corresponds to a power reduction ratio of almost exactly 2 : 1 since $0{\cdot}3$ is the log of $1{\cdot}995$.

The decibel is also used as a unit of sound, taking the normal threshold of hearing as a base, namely a sound intensity of 10^{-16} W/cm^2. The minimum increase in the loudness of a sound which the ear can detect is about 1 dB which represents an increase in acoustic power of 26%. Very faint cross-talk on a telephone line can be regarded as in the order of 30 dB down on the main signal, corresponding to a 1000:1 power difference.

Even when signal lines to an amplifier are screened, some high intensity radiation can have an adverse effect if the input signal is weak. Long transmission lines between an aerial or instrument and high-gain amplifiers usually require a pre-amplifier at the far end to improve the signal-to-interference ratio, since the pre-amplifier will strengthen the signal but not the interference picked up in the transmission line.

Radio frequency filters in the power supply lines are needed in many instances where the equipment is sensitive to voltage spikes caused by switching transients and the like. This type of interference may be fed into the mains from a radiated field, but is more likely to have been generated in the supply network itself.

A suitable earthing arrangement is one of the most important considerations in avoiding all types of interference, whether radiated or conducted, e.g. cross-talk, atmospherics or 'static', mains hum and the effects of transient switching voltages.

14.2.2 Electromagnetic compatibility (EMC)

The international forum for radio and general electromagnetic interference is CISPR (International Special Committee on Radio Interference). A Directive on the increasingly pressing subject of EMC was issued by the Council of European Communities in 1989 (Directive 89/336/EEC). This requires apparatus to be constructed so that

(*a*) the electromagnetic disturbance it generates is not great enough to prevent other apparatus from operating as intended
(*b*) and it has sufficient immunity to electromagnetic disturbance to enable it to operate as intended.

The apparatus envisaged under (*a*) is primarily where electricity is the source of power, and apparatus under (*b*) is primarily where low level signals are used and amplified. The growing need for conciliation here is being addressed by seeking to impose a degree of emission control on the one hand and a reasonable level of immunity on the other.

In accordance with the Common Market policy of trying to achieve a common market with reciprocal rules, each member state will have to measure the emission or immunity characteristics of its products by equivalent methods of test and evaluation when trading within the European Community. The development of the necessary testing standards has proved to be an awesome task for the technical committees concerned because each type of equipment requires an ad hoc test specification. For example, radiation depends on the aspect and height of the apparatus with respect to the ground plane and for portable equipment, the length and coiling of its supply flex will affect the test results. The emission from a vacuum cleaner may even be dependent on whether the dust bag is full or empty.

Testing and measurement standards for EMC, particularly as regards emission, are consequently detailed and technically complex although the critical field strength levels may be a straightforward commercial decision. It is the method of test and measurement rather than the actual results which are important. The necessary ongoing standards work is co-ordinated by the IEC Advisory Committee on EMC (ACEC) and it was intended to have sufficient harmonised standards in place by the end of 1993 to enable the EEC Directive on electromagnetic compatibility to be enforcible by the beginning of 1996.

It will be seen from the above that any set of results must be accompanied by a complete description of the equipment disposition during the test. The conditions of use must also be specified, e.g. maximum cable lengths, cable types, shielding and earthing. Such conditions also have to be incorporated in the user instructions. Furthermore, because a lot of interference is transmitted via the mains supply system, an artificial mains network having standardised impedance characteristics is necessary for the test.

Where emission from domestic equipment is concerned, the main criterion is its effect on nearby radio and television reception. The interference can be classified into continuous, such as might be produced by an electric motor or discharge lamp and discontinuous interference, such as from an electric fence or an electric iron. Complex items, such as washing machines, may produce both continuous and discontinuous interference. These days the dominant source of interference is often the portable transmitter-receiver, which operates normally at MHz frequencies.

For standard test purposes discontinuous emissions are generally related to clicks from switches and contacts. A click is defined in the Euronorm EN 50014: 1987 [BS 800: 1988] 'Limits and methods of measurement of radio interference characteristics of household electrical appliances, portable tools and similar electrical apparatus' as a disturbance which lasts more than 200 ms and which is separated from a subsequent disturbance by at least 200 ms. Obviously the more clicks/minute the greater the nuisance value. The permissible radiated field strength is therefore higher if the counted clicks during the test are below specified values.

When we come to EMC in industrial equipment, we are dealing not only with nuisance problems but also with matters of more importance, including safety. The international Electrotechnical Commission publication: IEC 801 Parts 1, 2 and 3 provides a common reference for susceptibility of industrial measurement and control equipment. Its recommendations have been adopted in the CENELEC harmonised documents HD 481 Parts 1, 2 and 3 respectively and the same wording has been embodied in the corresponding British Standards: BS 6667 Parts 1, 2 and 3, 1985*. These standards are primarily concerned with the sensitivity of industrial electronic apparatus to radiation from portable transceivers, as used by operating, servicing and security staff.

It is usually uneconomic to manufacture equipment resistant to every frequency or field strength which may conceivably exist. The severity of the interference

---

*BS 6667: Part 1: 1985
Electromagnetic compatibility for industrial-process measurement and control equipment General introduction
BS 6667: Part 2: 1985
Method of evaluating susceptibility to electrostatic discharge
BS 6667: Part 3: 1985
Method of evaluating susceptibility to radiated electromagnetic energy.

tests will therefore be in accordance with the environment in which the equipment is to be used and will be intended to demonstrate that the apparatus can function correctly when installed in its specified working conditions.

Because testing for compliance with EMC regulations requires expensive and complex equipment, plus a completely radiation-free environment, manufacturers can obtain permission to apply the Community mark (CE) to their products—assuming they comply with the other Community Directives—by what is termed the technical file route.

The file must contain

- A specification of the product
- An explanation of what it does
- A record of the means taken to meet the EMC Directive
- Data enabling a competent body to assess that it complies.

In the United Kingdom, a survey commissioned by the Department of Trade and Industry in 1989 emphasised that there would be a shortfall in the availability of engineers with the necessary skills and experience to deal with the growing demand for electromagnetic compatibility. The introduction of EMC as a discipline within the undergraduate syllabus was considered essential. Meanwhile, home-learning video-based courses have become more widely available.

Information technology, or IT, is taken to include all communication, process control, computer and data handling systems. In addition to IT for domestic, commercial and industrial purposes, other fields of application can be even more critical as regards safety. The subjects of medicine, railway operation and automobile management systems immediately come to mind as important in this context.

Meeting EMC specifications for medical purposes becomes more essential and more difficult with the developing intricacy and sensitivity of clinical equipment. Two of the main sources of interference in hospitals have been

(*a*) Surgical diathermy: during an incision, a low current, HF voltage is applied between the scalpel and patient. This cauterises the wound and prevents undue bleeding. Early diathermy generators produced such interference that they were collected and used in groups during the 1939–45 war in an attempt to jam enemy aircraft navigation signals. Besides creating the familiar herring-bone patterns on the old 405-line TV receivers, they could also conjure up random characters all over nearby computer screens.

(*b*) Staff location devices and bleepers, which can interfere with patient telemetering systems.

In a modern hospital, other contributions to the pollution of the electromagnetic environment can be from

- X-ray machines
- Body scanners
- Lifts
- Electric clocks
- Radio-frequency heaters
- Power transformers and air-break switchgear
- Fluorescent lamps

- Motors
- Radio transmitters and cell phones
- IT systems
- Pulse short-wave therapy
- ESD from synthetic textiles (a particular problem in operating theatres)

Magnetic fields from power cables and transformers can, of course, also erase or smudge data stored on magnetic discs.

The instruments most sensitive to radiation are those used to record small electric pulses generated by the body, the probes being externally applied electrodes. These include apparatus for electrocardiography (ECG), electroencephalography (EEG) and electromyography (EMG), detecting the natural functions of the heart, the brain and muscles respectively.

With respect to railway operation, signalling and train control have now moved from the mechanical to the electronic era, but the principles of lock and block are still essential for safe railway systems, i.e. interlocked signals and points to prevent conflicting movements and a headway of at least one block between movements. Track circuits, which are one of the fundamental IT systems for ensuring railway safety, are susceptible to various forms of electrical interference.

Ground currents from electric traction have been recognised and dealt with for over a century, but new problems are arising from harmonics generated by inverter drives on electrified routes. Because the equipment must fail safe, electronic monitoring devices are now being designed to interrupt the main traction current if interference limits are exceeded for more than a specified time.

Automobile management systems are subject to radiated interference but corrective action can in most cases be taken by the driver. The chief hazard is probably an accident due to distraction of the driver by a malfunction of some robotic control system. Compatibility between IT systems such as car phones and CB radios and an electronic car management system is clearly an important requirement for safety.

It seems probable that the increasing technical and commercial difficulties relating to EMC—not the least of which is the need for a single, reliable, 'clean' earth for each system—will be swept aside by a widespread adoption of optical wavelengths for data handling and transmission.

14.3 Electromagnetic pulse

To the military authorities, the letters EMP imply a nuclear electromagnetic pulse (NEMP) which is a brief but intense electromagnetic field caused by a nuclear explosion. It is comparable with the field from a lightning stroke, but the pulse from a nuclear detonation although intense is of shorter duration. Tests have shown that a wider pulse width is generally not as damaging to solid state devices as an increase in the peak voltage.

Prior to the 1960s, EMP was not really recognised as a technical problem, nor as a military threat. It was then noticed that screened instrument cables being used to record the results of nuclear test explosions would sometimes be fused by a detonation several kilometres distant. Since that time, the need to protect electronic equipment from the effects of transient radiation and electromagnetic pulse has given rise to specialised disciplines in protection and countermeasures which are collectively termed nuclear hardening. Good earthed screening is the obvious basic method but, as indicated above, the peak intensity of the pulse may be great enough to influence even screened circuits and components by direct penetration.

To consider EMP one needs first to understand a little about X-rays and gamma rays and their action on atoms—the most important as far as EMP is concerned being the Compton effect.

X-rays are electromagnetic radiations at very high frequencies, ranging from $0{\cdot}6 \times 10^{14}$ Hz (soft X-rays) to 3×10^{17} Hz (hard X-rays). These frequencies give them a wavelength of atomic dimensions so that they are capable of ionising atoms by knocking off their outer electrons. The theory of wave mechanics attributes this ability to a property of radiation which obeys the laws of momentum. The effective momentum can be measured and is directly proportional to the frequency.

While X-rays can produce free electrons, free electrons can produce X-rays. An X-ray machine consists of an electron gun, similar to that of a TV tube, which fires at a suitable metal target. As the electrons approach the charge fields of the target atoms, they become deflected. The change of direction corresponds to an acceleration of the invading electrons and, as we have seen, charge acceleration radiates electromagnetic energy. The frequency of this radiation can be selected within the X-ray spectrum by suitable choice of parameters.

Gamma rays are also high frequency electromagnetic radiations but are emitted spontaneously from disintegrating atoms. Their frequencies are a characteristic of the substance concerned and range from $7{\cdot}7 \times 10^{15}$ Hz for radioactinium to $6{\cdot}4 \times 10^{18}$ Hz for the radioactive thorium C″.

It can be seen that X-rays and gamma rays both have much the same range of frequencies. And whereas X-rays are generated by the deflection of near-miss electrons by atoms, gamma radiation is due to the disintegration of so-called radioactive atoms.

Gamma ray generation is usually accompanied by the emission of alpha and beta particles. An alpha particle is a helium atom which has forfeited both its outer electrons and so has a net positive charge of 2e. A beta particle is simply a very high speed electron emitted by a radioactive substance. We may note here that, being chemically hyperactive, alpha particles can scarcely penetrate the dead outer layer of skin so that materials which produce alpha particles are not hazardous unless breathed or ingested. Even a sheet of paper can stop all but the very highest energy alpha particles. Beta particles will penetrate a few millimetres of tissue and so are more hazardous.

Gamma rays correspond to hard X-rays, which can pass through the body and ionise atoms on the way by displacing electrons. Ionising radiation induces mutations and increases the risk of cancers. Generation of free electrons by such electromagnetic radiation is the Compton effect referred to above.

When a nuclear device—commonly known as an atomic bomb—is detonated, the core will consist mainly of a plasma of atomic nuclei, free electrons and electronic radiation in the form of light, X-rays and gamma rays. The yield from a nuclear explosion can be varied widely according to the design. In general, most of the energy released will be transmitted as blast and heat. The remainder is then nuclear radiation consisting chiefly of gamma and X-rays. These ionise the surrounding air molecules by displacement of their outer electrons. Because these electrons are much lighter and more mobile than their parent ions, which carry only an equal but opposite charge, they disperse far more quickly. In a fraction of a microsecond there is thus a rapidly expanding shell of electrons and a slower moving core of positive ions.

The gamma rays travel out from the point of detonation at the speed of light and reach their maximum flux density in a few nanoseconds (10^{-9} s). The radiation then continues at a lower level from the decay of radioactive elements produced by the detonation of the bomb. The electron and ion separation produces a very large transient field in response to the highly intense gamma and X radiation from the nuclear reaction.

If the charge separation were totally symmetrical about the detonation point there would be no resultant field. However, due to the proximity of the ground, the varying density of the atmosphere and the presence of the earth's magnetic field there will always be an appreciable resultant EMP. The ground below the explosion and the vertical gradient of the atmosphere causes the outward flow of electrons to be predominantly in a vertical direction. This creates, what is in effect, a diffuse dipole from which the EMP is radiated.

The gamma rays which are the prime cause of EMP can be divided into three time scales. The first wave consists of rays produced within a microsecond of the detonation. These are a direct result of the initial nuclear reaction. The second wave comprises those rays produced within one minute of detonation and are primarily from decay of atomic nuclei excited by the nuclear process when the bomb is detonated. The subsequent final gamma radiation is from radioactive residues with longer decay times.

Bearing in mind that the pulse is a single event caused by an increasing unidirectional current, its frequency spectrum is governed by the shape of the rise and decay of the pulse, as shown in Figure 14.9.

The harmonics accordingly extend from gigahertz (10^9 cycles/s) to virtually direct current in the tail. The EMP amplitude/time profile is similar to that of a lightning pulse and civil and military aircraft are able to use the same protective devices against both lightning and nuclear pulse.

Near the point of detonation, the whole of the frequency spectrum defining the profile will be present and the resulting peak within this area may create field strengths as high as 100 000 V/m. This can be compared with the highest average intensity considered safe for a human, which is about 200 V/m and a reasonably strong TV signal of 5 millivolts/m. The propagation of the HF content is in general limited to the line-of-sight, that is to say, as far as the horizon as seen from the elevation of the burst. Hence, beyond this distance, the rate of rise of the pulse will be somewhat flattened. Nevertheless, the longer wavelengths of the pulse can be detected over large areas of the earth due to the spherical space between the ionosphere and the ground acting as a waveguide.

As we have seen, EMP is mainly due to the X and gamma rays generated by the nuclear detonation. These react with molecules in the air and cause large Compton currents to be released outwards from the centre. If the burst occurs within the atmosphere, a fireball of about one to two kilometres diameter is thereby produced. If the bomb is detonated at an altitude which is above the atmosphere, the effect may not be so destructive on the ground but will create far more widespread interference with electronic and communication systems. Above the atmosphere the sideways X and gamma radiation will travel many kilometres before encountering any matter or molecules. In principle there will be no fireball

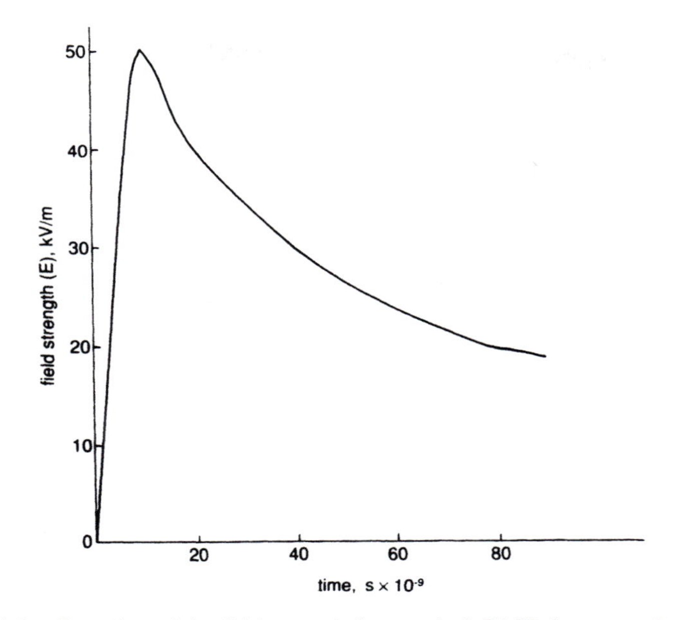

Figure 14.9 *Duration of the field strength in a typical EMP from a nuclear detonation. The peak value is reached in one hundredth of a microsecond*

although electrons will be generated in the atmosphere below the point of detonation — from horizon to horizon, i.e. over millions of square kilometres. They will produce more or less disruptive levels of electromagnetic radiation at ground level, in an area the size of Europe or China for example. As indicated by Figure 14.10, overhead power lines can also be affected.

Electrons from the nuclear blast itself will also be emitted, some travelling upwards to become trapped by the geomagnetic field. These electrons move in a helical path around the magnetic lines of force between north and south magnetic poles. Besides being reflected at the poles, they also precess from west to east so that the earth becomes enveloped in a sheath of electrons. Being in a helical path, they are technically accelerating and so will radiate and cause serious interference with HF communication channels.

This effect was demonstrated in 1962 by a 1·4 MT nuclear explosion above Johnston Island in the Pacific. High frequency radio communication between Hawaii and the Far East was interrupted for 6 hours and noticeable interference persisted for some 3 weeks.

14.3.1 EMP countermeasures

Because the duration of the peak of an EMP is so short the energy transmitted to an electronic system will be small but can nevertheless be destructive. The entry or coupling into electric and electronic systems may be

- Via an antenna or a conductor acting as such
- Directly through screening or metal enclosures

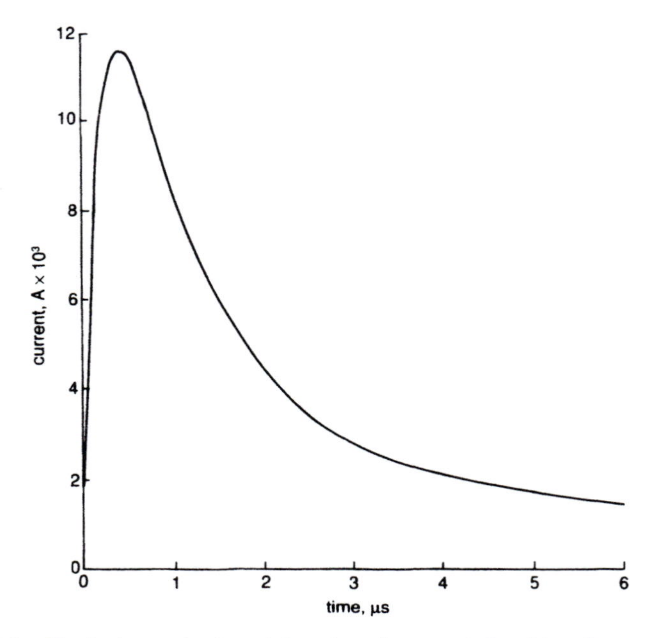

Figure 14.10 *Typical magnitude and duration of current induced in a long overhead power line by the EMP from a high altitude nuclear detonation. The magnitude will depend on the yield, distance of burst, length of line exposed to EMP etc*

- Through apertures such as doors, windows and vents in screened rooms
- Via cable and telephone lines etc.
- From re-radiation within a screened zone, by means of door handles and other conductors passing through the screen to act as a receiving dipole on one side and a transmitter on the other.

The ensuing damage can be

- Permanent, e.g. breakdown of semi-conductor junctions
- Temporary, e.g. lock-out (or in American English, latch-up)
- Single event upset (SEU), e.g. signal corruption.

Available countermeasures are briefly discussed below.

14.3.1.1 Screening or shielding

Totally closed steel cabinets can provide about 50 dB attenuation of the electric field strength E over a wide range of frequencies. Attenuation of the magnetic field H, however, is poor at lower frequencies.

14.3.1.2 Filters

These can be effective only if they discriminate and reject the frequencies in the EMP while accepting the working signal at an adequate level.

14.3.1.3 Voltage limiters

These can be connected across voltage-sensitive points in the circuit to protect them from over-voltages. Typical types use Zener avalanche diodes, non-linear resistors such as varistors, gas-breakdown surge arrestors and spark gaps.

14.3.1.4 Network hardening

This term implies the use of a refined signal code recognition circuit, able to disregard a random EMP signal and so avoid error.

14.3.1.5 Device hardening

Individually sensitive parts are selected or protected or duplicated so as to reduce the system vulnerability.

14.3.1.6 Circumvention

Rapid detection of EMP enables the system to shut off or transfer its operation during the period of interference. One means of achieving this is by the use of a saturable input transformer.

14.4 Radiological effects

All radiation can be harmful if it is above a certain field strength. The intensity can be measured in terms of volts per metre, which falls off as the distance from the source, or as watts per square metre, which decreases as the square of the distance from the source.

14.4.1 Heating by radio

Microwave cookers are designed to heat by electromagnetic radiation at 2450 MHz giving a radio wavelength in the oven of approximately 120 mm. At this frequency water molecules will vibrate in sympathy with the radiated energy. Their agitation raises the temperature of the food and acts as a load on the microwave generator. The radiation within the free space of the cooker will pass through non-conducting materials such as glass, ceramics, paper and some plastics, but will be reflected from conducting surfaces such as the oven walls and base and from metal dishes and cooking foils. The energy in a microwave cooker is produced by a magnetron self-oscillating valve connected to an antenna which illuminates the cooking zone. In general the energy is absorbed by the food up to a depth of 35 to 50 mm. To cook food with a greater cross-section than about 70 mm therefore requires some heat to reach the centre by thermal conduction. This is not a problem with liquids as they will mix by convection and can be stirred as necessary. For solid foods and when thawing frozen meals and the like, some extra time has to be allowed. A microwave cooker accordingly has several heat settings. On all but the maximum setting, the magnetron is switched on only for a proportion of the time. After each energised period there is a period while the temperature of the food is allowed to equalise towards the middle.

Even distribution of radiant energy also requires either the antenna or the food to revolve. Ovens without turntables are fitted with paddles behind a plastic partition. These reflect the microwaves back into the oven in different directions to create a more uniform wave pattern. Some ovens may be fitted with both turntables and rotating reflectors. Even so, those parts of the food which are near the corners of a container will have become more heated than the rest. This is because the radiant heat will have been received from more than one direction. For this reason microwave cooking should always be followed by a few minutes storage to provide a more uniform temperature and degree of cooking.

Because water molecules are heated by the magnetron frequency, the radiation is extremely hazardous—it can cook live tissue as readily as it cooks meat. Consequently all microwave cookers and warming cupboards must be reliably interlocked to ensure that the excitation is automatically stopped before the door can be opened. The European standard EN 60335-2-25 : 1991 Microwave ovens and the corresponding IEC Publication : 335-2-25 : 1988 specify a maximum permissible level of leakage radiation of 50 W/m^2, i.e. 5 mW/cm^2, at a distance of 50 mm or more from the surface of the appliance, with the door closed.

Accurate determination of leakage energy this close to the surface is difficult and requires special metering equipment. The instrument must respond only to radiation passing through a small defined area. Also, the distance from the source of the leakage will be very critical when the measurement is to be made at such close range. It can be compared to the difficulty of measuring very weak gas flows from a box containing a gas generator, without altering the flow of escaping gas.

Leakage from a microwave cooker is not self-evident and it is this fact which causes some apprehension in their use. It would be helpful if the energy generated by the magnetron were to be modulated in a way that could be readily detected by an ordinary commercial radio receiver.

14.4.2 Overhead power lines

At the other end of the frequency spectrum there has in recent years been some concern with regard to possible effects on mental and physical health of low frequency, electric and magnetic fields from overhead power lines. Extensive studies and reports have been reviewed by the World Health Organisation and by the National Radiological Protection Board in the UK. There was believed by some to be a relationship between a suspected increase in cancer mortality and exposure to low level electric and magnetic fields, but the results have not been particularly significant and none of the proposed hypothetical mechanisms has been found capable of experimental confirmation. In the USA, an influential lobby which believes that power lines are a risk to health is effectively forcing supply undertakings to put residential distribution cables underground. This of course has been strongly supported by environmental pressure groups.

14.4.3 Radio frequencies

There is in addition growing interest in medical effects of man-made electromagnetic radiation in general, partly as a result of national and international efforts to create an awareness of the need for electromagnetic compatibility. However, what is now termed electronic smog affects sensitive electrical apparatus rather than public health.

Portable transmitters are undoubtedly the most likely items to cause trouble and many poltergeist-like incidents have been reported as caused by mobile transmitters. Some are humorous, some are serious, such as a police car radio setting fire to the car while it was being filled with petrol. Meanwhile the Food and Drug Administration in the USA, although admitting that there is no proof that the devices are harmful to health, has advised users of car cellular phones to mount the antennae as far as possible from the drivers and their passengers.

An enquiry in the United Kingdom in 1992 found no evidence linking cancer to electromagnetic radiation emitted by TV and radio masts, overhead power lines, microwave ovens or TV and radio receivers. All these produce non-ionising radiation which is believed to be incapable of damaging living cells. On the other hand, as already mentioned, X-rays and higher frequencies and also ultra-violet radiation can damage living cells and cause them to mutate.

14.5 Bibliography

KING, G.J., *The practical aerial handbook*, 2nd Ed., Publ. Newnes-Butterworth, 1970

MESSENGER, G.C., and ASH, M.S., *The effects of radiation on electronic systems*, Publ. Van Nostrand Reinhold, NY, 1992

GHOSE, R.N., *EMP environment and system hardness design*, Publ. Interference Control Technologies Inc. Va, USA, 1984

BS 6527: 1988, *Specification for limits and methods of measurement of radio interference characteristics of information technology equipment*, Publ. BSI [identical to Euronorm EN 55022]

BS 800: 1988, *Specification for limits and methods of measurement of radio interference characteristics of household electrical appliances, portable tools and similar electrical appliances*, Publ. BSI [identical to Euronorm EN 55014]

BS EN 55011: 1991, *Specification for limits and methods of measurement of radio interference characteristics of industrial, scientific and medical (ISM) radio-frequency equipment*, Publ. BSI

KIDDLE, C., and MADDOCKS, T., *Management of electrical interference*: IT Infrastructure Library, 1992, Publ. HMSO

Chapter 15

Earth currents and their effects

15.1 Electric traction

As is well known, the direct current motor was invented many years before the alternating current induction motor. The d.c. motor goes back to the discovery by Faraday in 1821 of electromagnetic induction and rotation. His work was rapidly followed during the next few decades by a great variety of designs and patents for both motors and dynamos. Many of these incorporated novel and complex ideas. One of the early machines designed by Professors Elihu Thomson and Edwin J. Houston of Philadelphia included a brush-rocking device controlled by the load current. The commutator had only three segments and, to prevent damage to the surface, the sparks were blown out by synchronised blasts of air from an integral rotary vane pump as the gaps between the segments passed under the brushes.

But it was not until nearly the end of the nineteenth century that the possibility of using alternating current induction motors, without the need for commutators and brushes, was realised. This concept was independently suggested in 1887 and 1888 by two brilliant but little-known engineers: one was Galileo Ferraris 1847–1897 who became professor of industrial physics at Turin University. By theoretical reasoning he considered that it must be possible to cause a metal cylinder to rotate when placed between two sets of coils wound at right angles, each set being supplied with alternating current mutually 90° out of phase. He confirmed his theory by experiment. A somewhat similar device was announced the following year (1988) by Nikola Tesla 1856–1943, of New York.

High voltage direct current machines are not a practical proposition and with the possibility of high voltage transmission and utilisation and, at the same time, of avoiding the maintenance and replacement of brushes, the induction motor was of immediate interest for industrial drives and in particular for rail traction. Within the following few years the Allgemeine Elektrizitäts Gesellschaft, Berlin was actively developing a 3-phase, 50 Hz railway tramcar for a nominal speed of 200 km/h. The voltage proposed was 10 000 to 12 000 V, with transformers in the car supplying motors at 435 V; alternatively, a 50 000 V supply reduced by trackside transformers down to 300 V, with the traction motors wound for this voltage. Details of this work were given in a paper by O. Lasche of AEG presented at the 1901 International Engineering Congress in Glasgow.

Figures 15.1 and 15.2, respectively, reproduce two of the illustrations from this presentation.

The motors were of the wound rotor type, and although equipped with starting resistances, the vehicle was intended to run continuously at a fixed designed speed—presumably on a single dedicated route.

A 3-phase power supply for rail traction has since been commercially used in Italy, but with two phases overhead and the third phase in the running rails. However, this system has since been converted to direct current.

Until the comparatively recent use of variable frequency synthesis by electronic

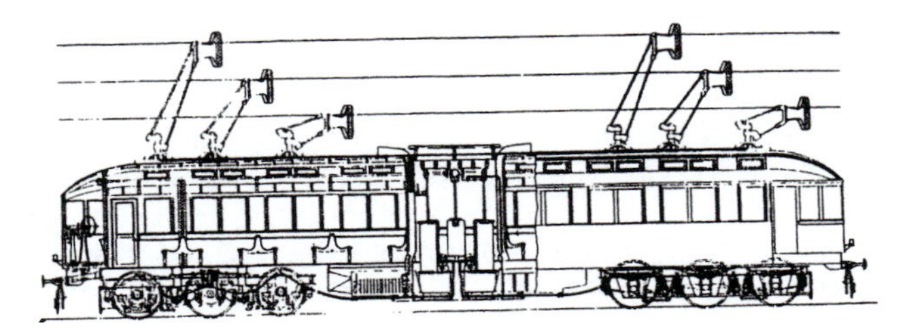

Figure 15.1 Cut-away side view of AEG 3-phase railcar

Journal of the IEE Vol. XXXI 1901–1902

Figure 15.2 Prototype AEG railcar with three-phase pantographs

Journal of the IEE Vol. XXXI 1901–1902

inverters, most traction motors were direct current machines, even when alternating current was collected. In this context it may be noted that an electrical machine with commutator and brushes, for all its cost and complexity, is an electrical analogue of an engine having a gear box with a continuously variable ratio. This feature, in conjunction with the series motor inverse torque/speed characteristic, offers a virtually unbeatable drive unit for traction purposes.

The sources of energy for electric traction can be

- Diesel-generator
- Rechargeable batteries
- Overhead line
- Third rail.

Power supplied by overhead line or third rail has the advantage of the higher generation efficiency of a central power station, together with a far better load factor, but the additional costs of power transmission need to be offset by using the highest practical voltage. There are very many voltages both a.c. and d.c. and several frequencies in use, but the de facto standard in Europe for long-haul electric traction has been set by the SNCF at 25 kV, 50 Hz. On-board rectification and control is now being superseded by electronic variable-frequency inverters feeding brushless induction motors. The resulting overall saving in weight and maintenance is quite profound. The motors can be totally enclosed, provided suitable cooling systems are included in the design and furthermore, a cage induction motor is much smaller and lighter than a direct current traction motor of the same power, typically, 1·8 t compared with 3·35 t.

Most line-fed power systems return the current via the running rails. Two notable exceptions to this are trolley buses and urban railway systems with a fourth rail for the return current. Although the running rails are normally of sufficient cross-section to take the full starting traction current, there is some volt drop incurred which cannot be ignored. A main line rake of coaches supplied at 750 V for example will send several thousand amps through the rails via the wheel tyres. This normally presents no problem on upstanding track, but the old horse-drawn tram tracks with rails recessed into the roadway often became filled with leaves and manure and the early 4-wheeled electric tramcars on these routes sometimes started with their chassis live to ground due to a high resistance between wheels and track, throwing off would-be passengers as they grasped the hand rail.

On railway and tramway systems most of the return current remains in the running rails but, as they are intended to be at earth potential, no specific insulation is placed between the rail and the ground—except in special circumstances referred to later. Consequently some of the traction current will flow out of the rails near the vehicles drawing power and return to the running rails in the vicinity of the nearest feeder pillars or sub-stations supplying the system. The depth and spread of these ground currents will vary according to the nature of the ground. Some of the current will take advantage of parallel metal services such as water mains where it may promote both external and internal corrosion.

At one time, the Board of Trade required the running rail voltages to be kept below seven and this involved the installation of rail drainage boosters, effectively to draw the current from the rail. Assuming that the return rail has a maximum voltage to earth at the tram, or train, it will leak most current at this point and the flow into the ground will decrease exponentially towards the region of the feeder cable where the fugitive current will be flowing back into the rail. In this respect the track can be regarded as a lossy return feeder.

Depending on the relationship between the rail conductivity, the ground conductivity and the resistance to the ground, from the rails to the track formation, the loss of current into the ground can vary quite widely and range up to about 30%. It will generally be greater—up to 50%—for a.c. return currents due to the inductive skin effect which causes the iron rail to have a far higher a.c. resistance than its d.c. resistance.

Electric traction on urban and city roads developed more quickly than railway electrification for a number of reasons, which can be considered later. By the 1920s every large town in the UK had its own electric trams backed by a highly developed specialised technology and a large indiginous manufacturing base which enjoyed a worldwide export market in similar systems. Among the earliest urban tramways was that at Blackpool, now one of the last remaining operating services. The sudden and general demise of the British urban tram has been entirely due to its incompatibility with the automobile on a shared road basis. Now, the motor car has so choked inner city life and commerce that new tramway systems, this time on segregated routes, are being constructed as rapidly as the original tramways were installed at the turn of the century.

The old urban routes created a number of hazards; there was always a danger to cyclists of catching their front wheels in the sunken rail and being dismounted. A further hazard for cyclists existed where the current collection was by means of a sub-surface arrangement. There were many types of these conduit systems, some of which were named after the city in which they were used. The general principle was that as the car proceeded it collected its current from an energised conductor fixed in a conduit or tube below the road, with a collector or plough carried below the car and passing through and along a continuous slot in the surface of the roadway. The armoured slot used by the London County Council was just wide enough to accept the front wheel of a sports bike, and so project the rider over the handlebars.

All open conduit systems suffered from ingress of water and dirt. And the eventual reversion of the 1884 conduit system in Blackpool to overhead collection was because the seafront route was at times flooded with sea water which choked the conduit with sand. At one time most major cities in the world had their own peculiar tramway conduit systems. Many of these were designed to work on a sealed principle, the section of exposed metal immediately beneath the car being made live by an electromagnet carried under the chassis. One of the simplest of these consisted of a hollow steel ball, which was drawn up by the action of an electromagnet carried under the car, to form a contact between the hidden continuous live conductor and an exposed sectional conductor. As the car proceeded, the ball rolled along with it and completed the circuit to the exposed section of conductor beneath the tram (see Figure 15.3).

Unfortunately, it was not entirely successful. There was no difficulty in making the ball follow the car while the magnet remained energised but, if the magnet lost power, the ball would drop and either roll away from the car, or the car would coast on, leaving the ball behind, in which case it was almost impossible to find the ball and re-establish the circuit. A large number of highly ingenious variations on this sealed conduit/magnetic actuation theme have in the past been tried and discarded.

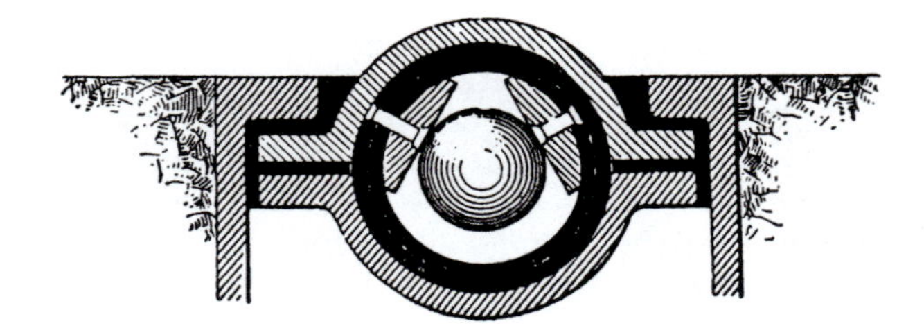

Figure 15.3 *Rolling steel ball raised by electromagnet under tramcar to energise section of conductor under the vehicle*

Modern Electric Practice, The Gresham Publishing Company, 1909

A major hazard for other vehicles was the notorious pinch where the rails converged towards the pavement to avoid a traffic island or due to a narrowing of the road. Perhaps the most dangerous of all, and the main reason why trams could not continue to cohabit with other vehicles, was the need they imposed on passengers to cross lines of moving traffic when entering or leaving tramcars in busy streets. The risk of falling under a tram was to some extent reduced by a wooden shovel at each end of the car and extending across its full width. This was arranged to drop down and scoop up a body before the wheels could dismember it. Further protection for other road users was given by powerful brakes, effected by magnetic shoes which clamped down on the rails, the power being generated by the motors of the tram as they were forced round by its momentum.

Overhead power lines for trams in built-up areas are obviously a serious hazard and the reason so many attempts were made to put the power supply below street levels was because many municipal authorities would not countenance naked overhead power conductors down busy streets crossed by thousands of telephone lines and other services. A number of safety devices were invented to guard against accidents resulting from fallen trolley wires or from other wires falling across the tramway supply. Most of these were mechanical linkages arranged to place an earth contact on the overhead line if it broke and lost tension, or if a line fell across it from above.

15.2 Railway signalling

It is the need for signalling and the provision of adequate stopping distances which primarily differentiates railways from tramways. Railway signalling and control must not only ensure safety but must be designed so that as far as possible, any error or malfunction still results in a safe situation – even if it brings the service to a halt. Over the years vital improvements have been made in the safety of railway operation, yet the work of the signalman is still a responsible task where human failure, or a single oversight can have the most dire consequences.

As Alan Earnshaw so poignantly relates in the preface to Volume V of a recent set of books on railway accidents

'. . . one can but feel sorry for men like signalman Sutton who stood in his box at Hawes Junction at 6 am on the morning of Christmas Eve, 1910, watching the sky turn dark red with the glow of a great fire: realising the result of his negligence he turned to a colleague and said "Will you go to Stationmaster Bence, and say that I am afraid I have wrecked the Scotch Express".'

Electricity as an aid to railway safety started with the electric telegraph and has continuously progressed to its present essential role in all fields of railway operation. The adoption of electric traction in place of steam or diesel engine power is of concern to the signal engineer. The reason is the possibility of interference between the very low voltages and currents used for signalling and telegraph circuits and the high voltages and currents required for electric tractive power. This has no doubt influenced many decisions against electrification, in spite of its manifest commercial benefits. Opposition has, of course, also been rooted in the pride of the traditional mechanical engineer. It is of interest to note that one of the inducements for suburban railway electrification was the success of municipal tramway undertakings in the commuter market.

The essential function of railway movement control is to ensure that trains are protected from danger due to following, converging or crossing movements of other rolling stock, locomotives and vehicles. The first safety rules used time intervals to segregate movements but it soon became necessary to adopt space intervals, the principle being known as block signalling which, like Pauli's exclusion principle regarding electron orbits, is intended to prevent more than one train being in a section of line, or block, at the same time. To achieve this, the distance between trains has to be carefully regulated. The length of a block can be varied to suit traffic conditions. At some places such as the approaches to a major terminus, it may be a few hundred metres; at others it may be several kilometres. In every case, however, the minimum distance to be observed depends upon the distance in which the heaviest and fastest trains can be brought to a stand on the gradients obtaining.

Block working became compulsory in the United Kingdom in 1892 following an Act of Parliament of 1889, although many railway companies were already using the system well before then. In fact, with the aid of the electric telegraph, the block system was in full operation between London and Dover by 1860. But the risk of human failure still remained. Having accepted a train into his section, the signalman might forget to reset the signal to danger while the train was in the section, so that a following train could then enter it. By a further refinement known as lock and block, the train itself puts the signal back to danger as it passes. In the 1870s, W.R. Sykes of the London, Chatham and Dover railway inaugurated this system whereby train wheels activate a treadle as they pass the signal, setting it to danger and switching an electric circuit to operate a lock in the corresponding signal box. Trains could thus be proved to have passed out of a preceding section with the signal behind automatically reset and locked at danger until the following section was clear. The remaining weakness was that, if the train became divided, the protective signal could still be released when only the front portion of the train had left the section.

This hazard has been overcome by what is termed track circuiting. The method is accredited to Dr. William Robinson of Brooklyn, N.J., whose first patent was dated 1872. It has been said that this invention has contributed more towards safety and the economic utilisation of railways than any other. Its great virtue is that it fails safe. That is to say, if the circuit is broken, or while there is any rolling stock on the line, the starting signal at the entrance to the section automatically resets to danger. It also enables mimic diagrams showing all train movements and positions to be presented at a central location.

The principle of operation is indicated in Figure 15.4. The electrical resistance of sleepers, ballast and track formation between a pair of running rails in a section is normally not less than 2 ohm and may often be considerably higher. An average value is a few ohms per kilometre. The rails at each end of the section are insulated from the rail ends of the adjacent sections and the regulating resistance is adjusted so that the track relay will drop out and hold the protecting signal at danger whenever the resistance between the rails in the section falls to 0·5 ohm or less. This sensitivity can usually be achieved with a permanent track circuit potential of about 2 V.

Thus we see that if the circuit is broken, or there is a vehicle on the line the relay is de-energised and drops out. This brings us to the question of electric traction and its effect on signalling. A third rail or overhead line power supply relies on a continuously conducting running rail for the current return, whereas track circuits require insulated sections. One solution is to use only one of the running rails for the return current. A more common solution is to use alternating current track circuiting with impedance bonds at the ends of each section. These will have a very low d.c. resistance of less than a milliohm, but a relatively high impedance to the track circuit frequency.

The use of concrete sleepers reduces the resistance between rails on longer sections below the requisite 2 ohms and, with modern high speed track employing concrete cross-ties, the rails have to be insulated from the concrete and the holding down clips are separated from the rail by non-conducting inserts. In some tunnels, notably in the Severn Tunnel, the dampness and resulting growth of algae cause the track circuit to have a resistance too low for practical purposes and in these circumstances an axle counting system has been tried.

Although the track circuit will fail safe in the case of a break in the relay circuit, including the removal of a rail, a possible unsafe, or wrong side malfunction could occur if the relay failed to drop out due to the presence of either an over-voltage or of a spurious voltage. Since only about 0·5 volts is sufficient to hold in the relay, it is clear that stray currents and stray ground voltages can be a serious problem. Where the main traction current in the rail is a.c., it is therefore necessary to ensure that the track circuit is impervious to induced voltages. Similarly, for

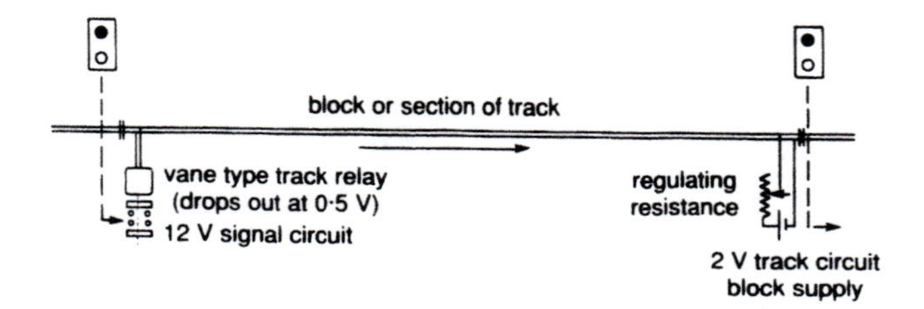

Figure 15.4 Principle and typical values of an elementary d.c. track circuit

direct current traction systems, alternating current track circuits have to be used. The relay contacts are of a material such as metal-impregnated carbon which will not weld or fuse together.

Modern intercity railways are rapidly changing over to centralised solid state signalling and routing, with radio contact between traffic controllers and train staff. In place of the traditional telegraph instruments, a lossy radio feeder is laid alongside the track so that continuous communication is possible. Track circuiting still demands immunity from ground currents and inductive interference, however. In addition, radio communication has to be compatible with the considerable radiated broadband interference from pantographs and third rail collectors.

15.3 Corrosion of buried structures

When studying corrosion one needs first to rehearse a little of the electrochemistry of metals. As we know, an atom consists of a nucleus surrounded by rings of orbiting electrons and the positive charge of the nucleus is exactly equal and opposite to the sum of the negative charges of its electronic planets. This is the normal or zero valency state of the atom, which is thus held together by electrostatic attraction balanced by centrifugal force, in much the same way as the solar system is stabilised by gravitation and centrifugal force. The outermost electrons have the weakest bond with their nucleus and so can most readily transfer their allegiance to the outer ring of other atoms. When an electron does this it leaves the original atom with an overall deficiency of negative charge. It is then no longer an atom but a positively charged unit known as a cation and the atom or group of atoms which wins the loose electron becomes negatively charged and is termed an anion, or more frequently when referring to chemical reactions both types of ion are called radicals.

Following the transferrred allegiance of an electron, the resulting cation and anion are mutually attracted and will combine to form a neutral molecule. In this way a metal can be united with another substance to create a compound such as an oxide, a hydroxide, a hydride or a salt. In the solid state most of these substances can be organised into a crystal or crystalline structure where all the positive and negative ions are bound closely together by electrostatic attraction. Crystalline here in its chemical context means a structure which is a uniform three-dimensional lattice or grid of repeated identical units. In this sense a wasp nest or an office block can have a crystalline structure.

If a compound of this type is dissolved in water it will either partly or completely dissociate into free anions and cations and in this state is said to be hydrolysed. The reason for the separation is because water has a dielectric constant or permittivity about 80 times greater than that of free space. Consequently, the attraction between the oppositely charged ions is reduced by this factor—allowing them to become more or less independently mobile. In this state the solution is a conductor, or electrolyte, and a potential applied between two electrodes in the liquid will cause counter migration of the positive and negative ions. The anions, carrying a negative charge, will flow towards the positive electrode or anode and the cations to the negative electrode or cathode. Although ions in solution are mobile, their migration velocity when subjected to a potential gradient is surprisingly slow. Even the hydrogen ion, which being the lightest is also the fastest, takes about 9 hours to travel between electrodes a metre apart with 100 V between them.

When a hydrogen atom loses its single electron, the remainder—usually written as H^+—is the most important ion in the field of electrochemistry as its concentration determines whether an electrolyte is acidic or alkaline. This property is normally specified as the $\log_{10}$ of the H^+ dilution expressed as the reciprocal of the concentration in gm-mols/litre*. It is referred to as the pH value where

$$pH = \log_{10}\frac{1}{[H^+]}$$

Water itself is only very slightly dissociated into H^+ and OH^- ions and the hydrogen ion concentration $[H^+]$ in pure water at normal temperatures is 10^{-7}, so that its pH value is 7. The concentration increases a little as the temperature rises, which means that the pH tends to fall. As can be seen, lower values of pH represent greater concentration of H^+. This denotes increased acidity. Rising values of pH indicate reducing acidity and increasing alkalinity. Since water is regarded as neutral—neither acidic nor alkaline—it follows that any neutral solution must also have a pH of 7.

Turning now to corrosion, it can be assumed that all forms of metallic corrosion, whether due to salt water, acid rain, bacteria or any other agency, are fundamentally electrochemical processes. The most obvious effect is the rusting of steel, which is a form of oxidation. We say a form of oxidation because, in the context of ionic or electrochemical activity, oxidation has a rather wider connotation than simply combination with oxygen. It means removal of electrons or the addition of a net positive charge. Thus, if a plate of metal such as platinum is placed in a solution which tends to oxidise it, the electrolyte will have removed electrons from the plate. The opposite to oxidation in this context is reduction, that is to say reduction of net positive charge. The oxidising or reducing power of a solution can be measured by the magnitude and sign of the potential given to the electrode.

Although we think of corrosion as primarily the production of an oxide, it can be a matter of oxidation, or reduction, or both. If no external voltage is applied there will be no external flow of current and the electrolytic corrosion must then consist of equal rates of oxidation and reduction reactions.

Consider first the oxidation process: this removes electrons from the surface of the metal and leaves soluble metal ions which can react with the electrolyte or adjacent substances. If the reaction forms an insoluble compound, such as an oxide, the corrosion product will accumulate to form a coating or deposit or a nodule at the site. If it is adherent to the surface and impervious to the electrolyte further action will be arrested. Some self-passivating films of this type can protect the surface even when they are only a few atoms in depth and may be extremely stable. A steel alloy containing 18% chromium and 8% nickel, for instance, forms a thin patina of chromium oxide which remains bright and prevents further chemical attack—hence the name 18:8 stainless steel.

With regard to the reduction process, this will take place in the absence of any external potential due to the electrons released by the oxidation process. Its nature

*A gram-molecule of a substance is the number of grams which is numerically equal to its molecular weight.

will depend on the oxygen content and on the pH value of the electrolyte—or in the case of buried structures, of the soil. In an acidic environment, which will have a relatively high $[H^+]$ value, the electrons released from the metal by the oxidation process will enable the H^+ cations to combine with dissolved oxygen to form water, in accordance with the equation

$$O_2 + 4H^+ + 4e^- = 2H_2O$$

If there is a shortage of dissolved oxygen, free hydrogen will be produced in accordance with the equation

$$2H^+ + 2e^- \doteq H_2$$

The presence of hydrogen can damage a number of metals by a process known as hydrogen embrittlement. In an alkaline environment, any dissolved oxygen will be able to combine with water to liberate the negatively charged hydroxyl radical, in accordance with the equation

$$O_2 + 2H_2O + 4e^- = 4OH^-$$

The presence of the alkaline radical OH^- can also weaken structures and other steel products, such as boiler tubes, by a comparable process known as caustic embrittlement.

It will be clear from the above brief description of the nature of corrosion that, as far as steelwork is concerned, there are two separate effects. Sites on the metal surface subjected to oxidation will lose metal as the neutral atoms become positive ions and pass into solution where they can be converted to compounds such as oxides and metal salts. And those sites subjected to the countervailing reduction effects will be liable to degradation by hydrogen or caustic embrittlement.

15.4 Cathodic protection

At those locations where an oxidation reaction takes place the metal attains a local positive potential and the effect is referred to as anodic corrosion. Reduction reactions are termed cathodic corrosion. If the electrolyte or soil is made positive with respect to the metal surface, the anodic corrosion will be reduced and the cathodic corrosion will increase. This is the principle of cathodic protection.

The voltage required will be quite low but the continuous current will be proportional to the area of metal surface to be protected. Cathodic protection by means of an impressed voltage involves a number of fairly detailed considerations together with appropriate testing of the final installation. The essential requirement is that the potential of the protected structure, with respect to the soil or water in contact with it, should at all points have a sufficiently negative value. But, because current will be flowing, this potential difference will vary from one location to another. Factors affecting these voltages will include

- The location of the anodes and their distances from the structure
- The electrical resistance between the various parts of the structure
- Differing surface conditions of the structure
- Non-uniformity in the resistivity of the ground between the anodes and the structure.

In many instances, the primary protection against corrosion is by a suitable non-conducting coating and cathodic protection is then installed to safeguard those parts where the coating may be damaged, porous or worn away.

As is well known, corrosion can be accelerated where two dissimilar metals are in contact in the same electrolyte. The metal with the more negative electrochemical potential will corrode more rapidly and, in so doing, protect the other metal. This is known as galvanic corrosion and the principle is used in the protection of steel by zinc coating—commonly termed galvanising. It is also applied in cathodic protection by means of sacrificial anodes. Here there is no impressed external voltage but anodes of a strongly electronegative material, such as magnesium, are buried or immersed near the protected structure and connected to it by insulated conductors.

When speaking of an electronegative material we mean a substance which appears negative to an externally connected voltmeter. But a so-called electronegative electrode is actually electropositive.

Thus, if the voltmeter reading in Figure 15.5*a* indicates that electrode B is negative with respect to electrode A, we say that the material of B is electronegative with respect to material A. In fact B is positive with respect to A because the source of the loop potential is within the electrolytic cell and is in the direction B to A.

It is electrode B which will tend to corrode by despatching its positive ions into solution and the mass of metal lost by corrosion can be exactly equated to a product of corrosion current and time. To suppress the corrosion of B we need to oppose this ionic current by making B negative with respect to its electrolyte—as indicated in Figure 15.5*b*. This current reversal can be obtained either by an impressed reverse potential gradient, or by the judicious placing of sacrificial anodes.

15.4.1 Safety considerations

There are a number of hazards associated with cathodic protection, some of which are not easy to see from first principles. Ships' hulls may be equipped with cathodic protection against the effect of sea water and the same applies to the steelwork and sheet piling of marine jetties. When a ship comes alongside, indeterminate currents can flow between the ship and the jetty. This is a particular hazard when

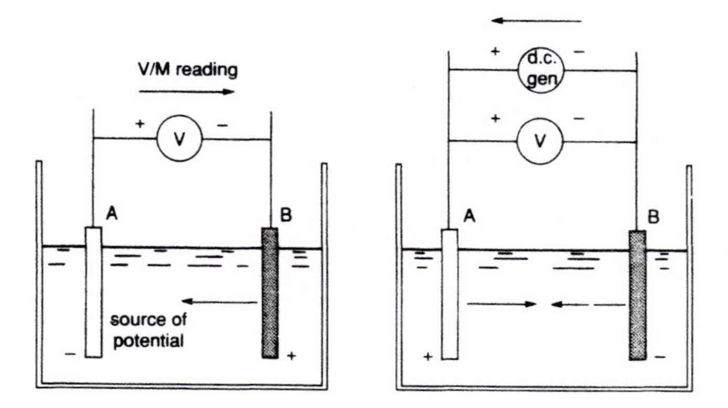

Figure 15.5 *a Direction of corrosion current for structure B*
b Direction of cathodic protection current for structure B

cargoes with flammable atmospheres are involved. Special electrical bonding arrangements are then necessary, in accordance with marine insurance classification society rules. Circulating currents should be allowed to flow only through an ad hoc low resistance connection and this circuit between ship and shore should be closed or opened only by means of a flameproof switch. Even after a solid bond has been made in this way, additional currents will still circulate through any parallel connections. Also, the magnitude of these secondary currents is little affected by the presence of even a very low resistance earth bond to the shore.

For this reason it is usual to connect every hose between ship and shore via an insulating section. Each cargo line should have only one insulated joint to avoid forming an isolated centre section which could become electrostatically charged. If both tanker and jetty are under their own active cathodic protection, it is nowadays considered advisable to keep the ship's structure unbonded. The hull is then electrically connected to the jetty through the water only.

Inspection and cleaning of any ship's hull or underwater equipment should not be attempted while cathodic protection with impressed current is switched on since even very low voltage gradients can be dangerous when the whole body is immersed.

In some circumstances cathodic protection itself can produce a flammable atmosphere. We noted above that the suppression of anodic corrosion tends to increase cathodic corrosion and that, in certain conditions, this will generate hydrogen gas. Where the interior of tanks is cathodically protected, there can thus be a danger of accumulation of a flammable atmosphere. Such tanks should be assumed to contain an explosive mixture until found to be safe by test. With sea water, other effects such as release of chlorine from the salt in the water are also possible.

Oil carriers normally have to travel one way in ballast, most of which can be pumped out while making passage through the calm seas of the tropics. The cathodic protection of the tanks therefore has to deal with at least three conditions; full cargo, full ballast and light ballast. For a 1000 m^2 tank wall the protection current required will depend on a range of factors and may be only 5 amps for a well coated tank, or over 100 amps for an uncoated tank in ballast. Although the impressed voltage is relatively low, the possibility of explosive atmospheres in lightly ballasted cargo tanks cannot be ignored. There will be an ignition hazard if the ullage falls below the bottom of the anodes.

With buried structures, as one might expect, we face quite a different set of problems. Many of these are peculiar to specific applications such as the protection of steel reinforcement in concrete and the protection of buried pipelines. These need not concern us here—of more relevance from a safety aspect are the interrelationships between cathodic protection and electric traction currents.

Where buried structures and services are near to d.c. traction systems, there is a source of ground current which is independent from any cathodic protection current. For rail services operated by direct current it is general practice for the running rails to be connected to the negative side of the power supply. As noted, some of the traction current will flow through the ground and some will take advantage of conducting paths provided by metal structures and services in the vicinity. Where the current is flowing from the soil into these local metal parts there will be reduced corrosion as this corresponds to the direction of flow with impressed cathodic protection. Where the metal parts veer away from the line of the rail track, and where they run close to the track feeder connection, the traction current will flow back into the ground and so to the rails or feeder. This will be in opposition to any imposed cathodic protection and will encourage the anodic corrosion.

To avoid the adverse affect it is helpful to bond local buried structures and parallel services such as water pipes and cable armouring to the return rails close to their point of connection with the sub-station, that is to say, at the point where the potential of the rail is at its most negative. This will ensure that the metalwork remains negative with respect to the ground throughout. Vagrant traction currents in nearby buried conductors are in this way induced to flow ohmically directly back to the return rails. The system is termed direct drainage and this will of course be additionally effective if the traction circuit incorporates its own drainage booster for the return current since this serves to depress the rail potential further below ground potential in the region of the sub-station.

Because of possible effects on track signalling circuits, direct drainage bonding between buried structures and services and an electrified rail system can be made only with the formal approval of the railway undertaking.

15.5 Bibliography

MORGAN, J.H., *Cathodic protection*, Publ. Leonard Hill [Books] Ltd. London: 1959.

ASHWORTH, V., BOOKER, C.J.L., *Cathodic protection*, Publ. Ellis Horwood Ltd., Chichester: 1986.

BS 7361 Part 1: 1991 *Cathodic protection Part 1 Code of practice for land and marine applications*, Publ. British Standards Institution.

Index

Acceptor resonance 162
ACEC 168
Acoustic shock 128, 133
Aerial 163–6, 173
Air discharge 129
Alpha particle 148, 171
Ammonium chloride 116–8
Ampere, A.-M. 144
Anaesthetic gas 154
Antenna, see Aerial
Arc voltage 122
Asbestos 125
Askarel 120–1
Attenuation 166–7

Ball lightning 127
Barrier device 83–5, 89
Battery ventilation 110–1, 115
Beryllium copper 125
Beta particle 148, 171
Blackpool tramway 181
Body resistance 26
Bonding 25, 32
Born, M. 159
Bowes, P. C. 94
Breakdown strength 144, 146
British standards
  dedicated 14, 17, 18, 26, 27, 61
  horizontal 11, 23, 54, 56, 61–3, 66–8, 72, 92, 96, 168
Brush discharge 130, 148–8
Buxton certificate 66–7

Cables
  insulation 12
  oxygen index 42
  smoke emission 42, 45–6
  types 43

Cadmium ignition 86–7
Capacitance effects 30, 118–9, 123, 141, 143
Centre-point earthing 19, 20
Circuit breaker, types 16–7
Circuit protective conductor (CPC) 16, 33
CISPR 167
Class II equipment 18, 103
Combustion energy 77
Compton effect 171
Conducting unit (c-u) 152
Contact resistance 104
Continuous grade release 54, 56
Corona discharge 130, 147–8
Cosmic rays 158–9
Coulomb, C.A. 144
Coulomb's law 144, 147
Crowbar circuit 22, 24
Crude oil 7, 152
Current balance protection 29, 31
Current limiting resistance (CLR) 84, 86, 88

Damping 161
Decibel 166–7
Diathermy 169
Dielectric constant, *see* Permittivity
Direct contact 24
Disasters 45–6, 52, 91–2, 100, 150, 153, 154
Displacement current 160, 165
Door-mounted equipment 102
Dust zones
  Zones Y and Z 96–7
  Zones 11 and 10 98

Earnshaw, A. 182
Earth current 39-40, 128 *et seq*
Earth electrode, see Earth rod
Earth fault clearance time 15-7
Earth-free environment 26
Earthing
  need for 18-21, 124, 126, 135-8, 146, 167
  reliability 22
  shortcomings 26-8
Earth leakage 27, 29, 30
Earth leakage breaker, *see* Residual current device
Earth loop impedance 15-7, 22-3, 25, 26, 27, 29, 33-8, 40
Earth rod 37-9, 77, 135-8
Edison battery (Ni/iron) 113-4
Electrical enclosure 18, 19, 173
Electrical fires 22-3, 27, 42-8, 56-9, 98-100, 111, 125
Electricity Supply Act 1926 31
Electricity Supply Regs. 1988 31-33, 36 40-1
Electricity supply systems 33-7
Electric shock 1-5, 99, 103, 119, 126, 143, 154, 155
Electrocution 2, 24, 99
Electromagnetic compatibility, *see* EMC
Electromagnetic spectrum 15-9
Electromagnetic wave 158
Electromotive force, *see* EMF
Electron 115, 123, 143, 165, 171-3, 185, 186, 187
Electronic system earth 136, 170
Electrostatic attraction 143, 155
Electrostatic discharge 128-9, 140, 142-3, 145-50, 153-4
Electrostatic field 144-147
Embedded temperature detector (ETD) 8, 10
EMC
  directive 167
  hospitals 169-70
  standards 168
  technical file route 169
  testing 168
  TIG welding 124
EMF 159-61
Equipotential zone 25
Execution by electrocution 2
Explosions
  dust 52, 91-8
  in mines 52, 63-4
Explosive gases, properties 56-9, 111

Exponential decay 149
Exposed-conducting-part 24, 32
Extraneous-conducting-part 25, 32

Factories Act 1, 2
Faraday, M. 157, 158, 166
Fatal accidents 23, 30, 50, 99
Ferraris, G. 178
Fibre optics 89, 170
Filled cable 74
Filtration, charge generation 151
Firedamp 68, 71
Flameproof enclosure 64, 65, 66-78, 184
Flameproof joint 71
Flammable paint 155
Flammable vapour 56
Flange gap 70
Fluorescent lighting 119, 121
Force between conductors 132
Fork lift truck, gas ignition hazards 60-1
Franklin, B. 133
French Automobile Club 114
Fuses
  limitations 13, 15
  types 16-7

Galvanic corrosion 188
Gamma ray 159, 171-2
Gas blanketing 153
Gas group 71
Gas ignition source 59
Gas ignition temperature 58, 60
Ghost image 165
Gilbert, W. 140, 141
Ground voltage gradient 136-7
Guarding 100-1

Half-life 149
Hands-on inspection 107
Harmonised standard 64-5
Hazardous area classification (HAC) 53-4
Heart 4-6
Henry, J. 157
Hertz, H. R. 157, 161
Hibbert, W. 114
Houston, E. J. 178
Humidity, effect of 146, 154

Ignition energy 59, 143
Immersion heater 104
Index of explosibility 93

Indirect contact 24
Inductance, effect of 85-6, 88, 123
Inerting 153
Infra-red radiation 126
Ingress protection (IP Code) 18-9
Insulation
  elastomeric 42
  limiting temperature 11-2, 43-5, 46, 48
  mineral (magnesium oxide) 46-7
  polyethylene 47-8
  PVC 44-6
  silicon rubber 47
  test method 43, 49-51
  thermoplastic 42
  XPLE, *see* polyethylene
Integrated circuit (IC) devices 146
Intrinsic safety 64, 79-90
Ion 115, 123, 185-7
Ionisation 147, 153
Ionosphere 163
Isolating transformer 101

Joule, J. P. 144
Jungner battery (Ni/cadmium) 112-3, 114-5

Kidd, A. L. 23

Langford, B. 94
Lasche, O. 178
Leclanché, G. 116
Lighting fittings 95, 105
Lightning
  bolt, *see* Ball lightning
  conductor 132-6, 138
  current 130
  electromagnetic interference 133
  leader 130
  rate of rise 131
  side flash 132-3
  voltage 131
  zone of protection 135
Lines of force 158, 161
Lithium battery cathodes 115-6
Lithium, properties 115
Lower explosive limit (LEL) 53, 56, 58

Magnetic field 157-8
Magnetic flux 156, 159, 160
Magnetomotive force (MMF) 158
Magnetron 175, 176
Manual metal arc (MMA) welding 122-3, 126
Maximum explosive safe gap (MESG) 69-70
Maxwell, J. C. 157-9
Metal powder 91
Microchip 145
Microwave cooker 175-6, 177
Mordey, W. M. 40
Motor torque-speed curve 7, 10-1, 180

National Electrical Code (NEC) 54
New Approach Directive (CENELEC)
  conformity category 65
  essential safety requirement (ESR) 65
NICEIC 103
Notch filter 166
Nuclear hardening 171, 174-5

Oersted, H. C. 157
Operating theatre 154
Overhead line, HV 5

Particle ignition 72, 81
Peak pressure measurement 68
PEN conductor 33-5
Permeability 160
Permittivity 120, 160-1
Personal protective equipment (PPE) 110, 112, 127
Phillips, H. 69
pH value 186
Plante, G. 109
Plug top (BS 1363) 18, 27, 103
Polarisation 117
Polychlorinated biphenyl 120-1
Portable transmitter 177
Potassium hydroxide 112, 114
Potential equalisation conductor (PE) 33, 102, 103
Pre-amplifier 167
Pre-compression 68-9
Pressure vent 110
Primary grade release 54, 56
Prospective short-circuit current (PSCC) 13, 32, 49
Protective multiple earthing (PME) 35, 37, 40-1
Pumping rate 151
Pump, torque-speed curve 7, 10-1

Q-factor 161, 162

Rack-mounted equipment 102
Radio-active ionisation 153

Radio spectrum 165
Radio waves 158-9, 165
Rail drainage 180, 190
Railway operation 170, 178-85
Range of flammability 57
Rate of rise 131, 172
Rejector resonance 162
Relaxation of liquid 152
Residual current device (RCD) 28-32
Resonance 161
Restriking voltage 85
Resuscitation 4
Robinson, W. 183
Rupturing capacity 13, 20

Safety device 100-2
Safety extra low voltage (SELV) 7
Sal ammoniac, *see* Ammonium chloride
Screening 104, 166-7, 170, 173-4
Secondary grade release 55-6
Severn tunnel 184
Shielding, *see* Screening
Short-circuit energy 77
Shut-down inspection 107
Skin effect 181
Skin resistance 3
Socket (BS 1363) 18
Soil resistance 37-40, 187
Spark energy 85
Sparking test 85
Splash filling 151
Spray gun 155
Stagnant gas condition 55
Starting battery 109, 112
Static eliminator 147-8
Static potential 141-2, 144-5
Static shock 143
Sulphur 91, 94
Sulphuric acid 110
Surface temperature 79, 95
Switchroom 104-5
Sykes, W. R. 183

Tesla, N. 178
Thermal lagging 56, 94-5
Thermistor 10
Thomson, E. 178
Time constant 149
Track circuit 183-5
Tramway systems 180-2
Transformer 104
Tungsten inert gas (TIG) welding 124

Ultra-violet radiation 126, 177
Un-earthed system 19
Uninterruptible power supply (UPS) 111
Upper explosive limit (UEL) 53
US Bureau of Mines 92-3

Ventilation 110-1, 115, 127
Ventricular fibrillation 4, 29-30
Visible light 158-9, 165
Visual inspection 105-6
Volta, A. 108, 144
Voltage elevation 32
Volta's pile 108
Volume resistivity 148

Welding electrodes 122, 127
Windings
 overheating of 9, 25
 protection of 10, 11
Windsor Castle 100
Wiring Regulations BS 7671 23, 25, 28, 30, 33, 40, 60
Work in confined spaces 127

X-rays 159, 171-2, 177

Zones of gas hazard (2, 1 and 0) 53